PRACTICAL AV/VIDEO BUDGETING

by Richard E. Van Deusen

Knowledge Industry Publications, Inc.
White Plains, NY and London

Video Bookshelf

Practical AV/Video Budgeting

Library of Congress Cataloging in Publication Data

Van Deusen, Richard E.
   Practical AV/video budgeting

   (Video bookshelf)
   1. Instructional materials centers—Accounting.
I. Title.   II. Series.
HF5686.I53V36   1984   001.55′3′0681   84-17079
ISBN 0-86729-100-1

Printed in the United States of America

10   9   8   7   6   5   4   3   2   1

Table of Contents

List of Tables and Figures

List of Budgeting Forms

Foreword

WARNING! This book is not intended to make you an instant accounting expert. In fact, I would just as soon accountants didn't read it at all. For one thing, they would be able to punch it full of holes, as I take a number of liberties with standard accounting practices and procedures. Also, they would discover some of the secrets of how we audio visual professionals manage to circumvent a lot of those practices and procedures.

There are a couple of things to remember about accountants. First, financial statements are written so as to be understandable only to other accountants. This is a form of job security. (Admittedly, we in the audio visual profession have our own ways of trying to find job security as well, so we can't knock accountants for doing it their way.) Second, while accountants simply cannot believe that audio visual people can add two and two correctly, they have no trouble believing that accountants can be experts on the audio visual production process, especially the costs.

I should say that I have nothing personal against accountants—both my father and brother chose that profession. Furthermore, in spite of my protestations that I am "creative," I frequently have to don my green eyeshade, pick up my quill pen and attend to the essential bookkeeping side of our business.

And it is a business. Whether you are an independent producer, work for a commercial production company or are part of an in-house audio visual production staff, you must be concerned with the process of budgeting. Budgeting is also important whether you are a manager, producer, director, writer, engineer or production assistant—anyone involved in designing, developing or using the audio visual media.

To simplify the discussion in this book, I use the term "company" to refer to the organization to which the AV function is attached. This blanket term encompasses all types of parent organizations: corporations, government and nonprofit agencies, health and educational institutions, etc. I use the term "manager" to refer to the individual responsible for managing the budget, whether that person has the rank of manager or is a producer, project manager or whatever. The term "department" refers to the audio visual function, although it may in fact be a division, a unit, a section or even just an individual.

This book is directed primarily, however, to the audio visual manager/producer working in a corporate environment. That is, after all, my personal experience, as well as the experience of most of the people with whom I regularly deal. By implication, however, much of what is included will be applicable to those working for other types of organizations.

The primary purpose of the book is to help the AV professional to understand the need for sound accounting and financial data and to be able to use it in meeting the primary objective of producing cost-effective audio visual media. The book is also intended to help the creative mind deal effectively with company managers, clients and, yes, accountants, who still must approve the bills to pay for what we do.

The information in this book comes from my own 20 years of audio visual management experience with a large financial services company. During that time, I have had many opportunities to learn from my fellow professionals in other corporate, government and educational institutions, as well as from commercial producers and, of course, the financial experts in my own company. While these people are too numerous to mention, I owe them a great debt for all the things I have learned through our ongoing exchange of ideas and mutual problem solving. Much of this has been on an informal basis but was made possible through organizations such as the International Television Association (ITVA) and the Audio Visual Management Association (AVMA).

Specifically for this book, I conducted considerable additional research, most notably a survey of the AVMA membership. The results of that survey are given in Appendix D and discussed, where appropriate, throughout the book. I am indebted to the many members of AVMA who willingly shared information, ideas and materials with me for this book.

The reader will find throughout this book a number of case histories in the name of the fictional Pimlico Corporation. The fact of the matter is that while my colleagues were quite willing to relate their trials and tribulations off the record, few were willing to let me quote them directly or attribute the source. It would appear that most corporations are not yet willing to tell their stockholders just how much they spend on audio visual support—another sign of the suspicion with which the function is still viewed by a lot of corporate financial managers.

I am indebted to the many people who provided material or information quoted or paraphrased in the book and whose names appear with those specific references. I am also grateful to Bill Hight of the AT&T Resource Management Corp. in Lisle, IL, who suggested that I write this book and with whom I have shared the podium at ITVA seminars on budgeting.

I must also express my thanks to Richard J. Goff, who has been my boss for the past five years and who taught me a lot about how the corporate mind works, as well as about budgeting, financial management and creative accounting. Much of what is here came from him, directly or indirectly.

Finally, I am grateful to my son Derek for his patience and indulgence in introducing me to the world of computers and for his efforts in retyping the manuscript of this book.

Richard E. Van Deusen
July 1984

*In business, reputations are
made on what you spend,
not what you save.*
Anonymous

1 Overview of the Budgeting Process

Accountants define a budget as "a plan expressed in financial terms." That may seem simplistic, but it is in fact the essence of the budgeting process. The plan begins with a definition of the work to be done in the specified period of time and then gives a determination of the cost of performing that work, including the costs for people, floor space, equipment and outside services.

Without the element of the plan, the budget becomes meaningless, and the audio visual manager is hard put to gain approval, as many of us know from sad experience. Furthermore, the plan must be more than just a pie-in-the-sky projection of things we would like to happen—it must be specific, have definable objectives and be attainable. There is one more essential element: the plan must support or further the overall objectives of the corporate organization.

THE START-UP BUDGET

There are two planning scenarios, which differ slightly. The first, and less common, is the "starting from scratch" plan. In this case, the AV manager is asked to develop an entire audio visual activity where none currently exists. It may be as extensive as the creation of a whole new department, or it may be the development of new services to meet new needs, such as introducing a new line of business or absorbing a new subsidiary. In this case, it is necessary first to establish the needs, then to determine what resources are required to meet those needs and finally to put a price tag on the cost of providing those resources.

1

THE ANNUAL BUDGET

The second, and much more common, scenario is the annual process of budgeting for an ongoing AV activity in an organization. It is the routine chore of annually projecting the work that will have to be done in the succeeding fiscal period and then comparing the cost of doing that work with the cost of doing a comparable amount of work in the current or preceding fiscal period. Simply put, you look at how much work you have to do next year as compared to how much work you did this year, and then see whether you will need more or less money to do it.

Budget Obstacles

In this case, you will start with a base figure that includes the current costs of salaries, floor rental, equipment depreciation and other operating overhead as well as the cost of buying needed services on the outside. If the work projected for the next year will be less, it may be possible to meet those needs with a smaller staff, less space and certainly less expense in buying services on the outside. However, the problem in this situation is that you cannot turn some of the expenses on and off like a tap.

Fixed Costs

True, you can fire or lay off staff, but that is an extreme that most companies avoid except in the face of major expense problems. Getting rid of space is not much easier. If you lose staff, you might be able to get rid of some office space. However, reducing the size of fully constructed studios or photo labs will cost much more than can possibly be saved in space rental charges, even if another tenant is willing to move in.

Equipment costs, too, can be difficult to control. You cannot cancel leases and restart them at will without suffering a substantial penalty. Owned equipment can be sold, but usually only at a fraction of its currently depreciated value; you then wind up having to carry the difference in depreciation value on your books. Putting equipment in storage also does not wipe out the depreciation charges.

Company-Dictated Cost Increases

Then there are company-dictated cost increases over which the AV manager may have little control. In most organizations, annual salary increases are established by the company. Even where they are not, the AV manager is going to have a morale problem if no one on the staff gets a salary increase for an extended period of time, especially if the staff has to work harder or longer hours. The company may also raise the rent rate or other mandated costs, such as employee benefits, or the charge for other internal services.

The Impact of Inflation

Outside services and supplies will be subject to inflationary increases over which you have no control. The cost of tape—both audio and video—increased by 20% or more during the oil embargoes of the 1970s, causing many AV managers to exceed their budgets for duplication even though they did no more work.

With even nominal inflationary factors at work, to keep the annual budget the same as the preceding year amounts to a budget decrease. This is something that faced 12% of audio visual managers in 1983. For 1984, 24% had to deal with a similar situation. In addition, 21% faced actual budget decreases in 1983, and 16% had their budgets reduced for 1984. Thus, we find that one-third of the audio visual managers in industry had to find ways of reducing expenses in 1983. The situation deteriorated in 1984, when fully 40% of the audio visual managers found themselves having to find ways to cut costs—presumably while still providing some semblance of their previous level of service. In the case of actual decreases in budgets, the average reduction amounted to 11% in 1983 and 15% in 1984.*

Budget Constraints

Ultimately, then, the final approved budget may seem to have little relevance to the plan. It is not uncommon for the audio visual manager to tell his or her superiors that an increase of 20%, including new staff, new space and new equipment, will be required to meet the needs of customers in the next year, only to be told that the budget constraint will be to hold expenses to a zero increase.

The reason this happens is obvious. Most corporate budgeting begins at the bottom, with projected estimates submitted by each department, and proceeds up through the chain of command to the decision-making level. Each department projects its income as well as its expenses. At the decision-making level, all income and expenses are compared, and then expenses are reduced in accordance with anticipated income. (I say reduced since, to my knowledge, there has never been a situation where the decision-makers decided that expenses weren't high enough.)

Contribution to Profit

Normally, not all departments get equal treatment with regard to reducing or holding the line on budgets. Those that contribute most to the profitability of the company will be given the greatest leeway in budgeting—wherein lies a major problem for AV. The audio visual activity is competing for budget dollars with the sales department, manufacturing, engineering and a host of others, including accounting. To borrow the phrase of an executive of my acquaintance, "Audio visual is a foreign enterprise within the corporation." That is, it does not contribute directly to company profitability. Furthermore, it is difficult to come up with concrete examples of ways in which the function makes even an indirect contribution.

---

*This information is from the survey of the Audio Visual Management Association conducted by the author in 1983. For a full discussion of the survey, see Appendix D.

To say that having our own internal facilities and staff saves money because we can provide services at less cost than purchasing those services on the outside begs the question of whether the services are in any way essential. The fact is, a company can survive quite nicely without any audio visual support at all, regardless of what we, as audio visual professionals, may think. Of course, there are specific instances that clearly demonstrate the value of the audio visual media, but they are few and far between. Many audio visual managers feel they are living with Edgar Allan Poe's "The Pit and The Pendulum," with the pendulum of cost control slicing inexorably away whenever company profits go into a slump.

Lack of Understanding

To make the situation appear even more bleak, audio visual has the misfortune to be both relatively new to the corporate world and highly visible. Because it is new, there are a great many executives in positions of responsibility who have not yet been convinced that there is any value at all to audio visual communications. We are not the first to have been faced with that problem. It is well known that advertising departments had to contend with a similar lack of understanding, and even the telephone people were probably in a similar situation at one time.

It is interesting to note that, in those companies where audio visual has come to occupy an important and central position, the chief executive officers both believe in and make full use of the communications capability of the media. This tends to spread from the top down and thus throughout the organization. Executives like James Burke of Johnson & Johnson and Robert Beck of Prudential understand what the audio visual media can do to extend their influence throughout their organizations. As a result, those AV departments have a respected and solid position within their companies. (Unfortunately, unless that attitude spreads to the rest of the organization, the retirement of such an executive can have a devastating effect.)

Visibility

Part of the problem of AV's visibility has to do with the equipment—it is both expensive and obtrusive. A photographer covering a company reception with a motor-drive Hasselblad and strobe light is pretty hard to ignore. It's even harder to overlook the portable television camera, complete with full crew and lights, covering the annual meeting. Is there an audio visual manager anywhere who has not been asked at one time or another, "I like what you're doing, but do you really have to have all that equipment to do it?"

Is it any wonder that when it comes time to cut budgets, executive minds tend to turn toward the operating areas which they don't understand very well and which appear to be spending a lot of money doing whatever it is they do?

RECENT AV BUDGET TRENDS

While a substantial number of audio visual departments faced budget reductions in 1983 and 1984, remember that a majority had budget increases. In fact, 67% had increased budgets in 1983, and 60% were increased for 1984. More often than not, the same organizations got increases in both years, just as the same organizations that had their budgets reduced in 1983 had further reductions in 1984.

Furthermore, budget decreases appeared to result more from the economic pressures of the recession and the subsequent slow recovery than from any failure of management to appreciate the value of AV or of AV managers to present the value of the function properly.

Table 1.1 summarizes overall budget projections—both increases and decreases— for 1983 and 1984. A budget increase of 5% or less can be considered as staying even with inflation during the two-year period. It is significant that even in 1984 a majority of all companies increased their audio visual budgets by more than just the rate of inflation. Table 1.2 shows the average percent change in the budgets for the two years.

The small size of the sample (69 responses from an estimated 1200 major corporate AV departments) makes it difficult to determine whether the budget decreases noted have any longer term meaning. Only time will tell. However, I am

Table 1.1: AV Departments with Budget Changes, 1983-1984

	1983	1984
Budget Increases		
More than 5%	61%	53%
5% or less	6%	7%
Total	67%	60%
Budget decreases		
Actual decrease	21%	16%
No change*	12%	24%
Total	33%	40%

* No change classified as equivalent to a budget reduction.
Source: Audio Visual Management Association survey

Table 1.2: Average Percent Change in AV Department Budgets, 1983-1984

	Increase	Decrease
1983	13%	11%
1984	12%	15%

Source: AVMA survey

not aware that any company audio visual function was completely eliminated during 1983 (though some—for example, International Harvester—came pretty close). I do know that several were closed in the early 1970s, though most of them were later resurrected.

THE BUDGETING PROCESS: HOW-TO AND HOW-NOT-TO

To get a better idea of just how the budgeting process works, let's look at a hypothetical but typical situation. The audio visual communications unit in the public relations department of the Pimlico Corporation has a staff of eight, including the manager. The unit produces slides, does a lot of still photography and an occasional motion picture, and produces ¾-inch videotapes in a small studio. They provide services for the personnel training department (mostly videotapes), the marketing department (some videotapes and an occasional film, but mostly slides and photography for sales brochures and exhibits) and their own department, public relations.

Budgeting at Pimlico

The annual budget for the current fiscal year totals $756,200. Of this amount, $456,200 is for internal expenses, including salaries, rent, depreciation, etc., and $300,000 has been budgeted for outside services and materials.

To establish the budget for the succeeding year (the company operates on a calendar year basis), the manager must contact all of the users and potential users of audio visual services in August so as to be able to develop budget figures in September. The timetable calls for submitting the budget figures to departmental management in October. The AV budget is combined with the rest of the PR department's budget to be presented to the corporate finance committee for an initial review. Final approval comes in late November.

At Pimlico, the cost allocation system for AV services is full overhead. That is, the AV budget covers all expenses for all clients; clients are not assessed for any part of the services used. The cost is borne as company overhead and allocated back to income producing departments as operating overhead, as is typical of all service organizations in this company. (See Chapter 2 for a full discussion of cost allocation systems.)

Gathering Information

Of course, summertime is about the worst time to try to get projections of client needs for the next year. First, most of the clients have not yet decided what they are going to do in the fall, much less what they might want next year. Furthermore, vacation schedules make it very hard to find the right people to talk to.

But by persisting (and postponing his own vacation), the manager is able to get a pretty good idea of what his clients are going to expect of him next year. Most will want pretty much what they had for the current year. However, the marketing department is planning to try to crack a new market for one of their products and wants to produce new sales brochures plus two motion pictures for point-of-sale purposes. The personnel department is also going to introduce a new career development program and will want to do two videotapes.

Calculating Costs

Having gathered this information, the manager now closets himself with his adding machine and goes through the process of calculating what this additional work will cost. At about the same time, the company accounting department comes through with its projections on expense increases for the next year. Salaries are expected to increase an average of 7%, which is not unusual. However, the floor rent is being raised from $20 to $21 a square foot, telephone expenses are expected to increase by 10%, and a new employee benefit package is being installed that will raise the benefit factor from 35% of salaries to 40%. The AV manager knows he can figure outside expenses and travel on the basis of an 8% inflation factor. In addition, $10,000 worth of new equipment is needed, adding to the depreciation budget.

Without even considering the additional work, he produces the initial budget comparison shown in Table 1.3. The manager is facing an 8% budget increase just to continue doing the same amount of work. But our manager believes it will be possible to produce the additional work without adding to the staff. This will probably require some overtime, and his nonexempt staff must be paid for overtime. He estimates that would increase the salary category by about $4000. In addition, the films will require extensive travel, so he adds $6000 to the travel budget. The external expense line is also affected. There will be about $8000 worth of talent charges for the personnel videotapes and about $20,000 in additional materials charges for both the films and the tapes.

To accommodate all of this additional work, the budget will have to be raised to $856,160, or a total increase of over 13%.

Table 1.3: Initial AV Budget Comparison, Pimlico Corp.

	Current	Projected
Salaries	$200,000	$214,000
Employee benefits	70,000	85,600
Space rent	100,000	105,000
Telephone	9,200	10,120
Travel	18,000	19,440
Membership dues	500	500
Books/periodicals	500	500
Equipment depreciation	25,000	26,000
Maintenance	18,000	18,000
Miscellaneous	15,000	15,000
External expenses	300,000	324,000
Totals	$756,200	$818,160

Preparing the Proposal(s)

Having accumulated the financial data, the next step is to prepare a written budget proposal for his departmental management. This includes not only the expense item comparisons, but also a detailed report on just what work is to be done, particularly the additional projects. He drops the entire package on his vice president's desk on a Monday morning in early October, after having spent a good part of the weekend rechecking the calculations. Late in the afternoon, he is called into the vice president's office to talk about the budget, an event that will cause him to miss his car pool home.

The first question his vice president hits him with is, "Why do the marketing people need two films? Wouldn't one be enough?" At this point the manager knows it will be a long afternoon. By 6 PM, the manager has answered all the questions but the first one. So, on Tuesday morning, he contacts the marketing department to dig out the answers, including a very plausible explanation as to why the marketing people need two films. This process takes until Friday, at which point he meets with the vice president again to report his findings.

While the vice president does seem to accept the report, his final comment (again at 6 PM) is, "OK, that sounds reasonable, but you'd better rewrite this budget proposal for me. Oh, and I need it Monday morning, as I've got a preliminary meeting with the finance committee Monday afternoon." After another weekend spent closeted in his den, the manager is able to drop off the new proposal late Monday morning. The vice president's comment is that he hopes it will go through but, "Expenses are going to be a little tight next year."

The next morning, after another call to the vice president's office, the manager is told flatly, "As I expected, expenses will be tight next year and additional funding is needed for the research area, so we're going to be held to a 5% increase on audio visual expenses." Then he adds, "Oh, that will be 5% over this year's actual, not budget."

Other than the implications of additional weekends and evenings spent poised over the calculator, the concern now is how to reduce an $856,000 budget to a $794,000 budget. Not only that, based upon the actual work that has been done so far this year, it seems that the actual expenses will come in about 6% under budget, or about $710,000. Next year's budget will have to be pared down to $745,500, or less than had been budgeted for the current year.

CONCLUSION

The situation is not unusual, and it's certainly not unique to AV. How it is resolved is what this book is all about. We will deal with how to avoid this kind of Catch-22 situation in the first place—and, failing that, how to extricate yourself once you're in. The essence of the solution is a well-defined plan and a good system of fiscal controls.

There is one saying that the frustrated AV manager might remember, the source of which I don't recall: "Nobody ever got promoted for coming in under budget."

2 Cost Allocation Systems

There are almost as many different cost allocation systems as there are accountants to dream them up. However, in general, all systems can be divided into five primary types.

THE FIVE TYPES OF COST ALLOCATION SYSTEMS

I came up with the following categories some years ago, and they seem to have borne up with time:

- Full overhead

- Direct cost allocation

- Full cost allocation

- Full chargeback/cost center

- Full chargeback/profit center

These categories distinguish among the ways of allocating both indirect costs (salaries, rent, depreciation and other "overhead" expenses) and direct costs (payments made directly to outside suppliers). The names are more descriptive than technically correct from an accounting standpoint. However, they do provide some point of reference.

While, by definition, they are methods of cost allocation, they also provide the basis for establishing and administering operating budgets. Here, then, are more detailed explanations of each.

Full Overhead

The term "full overhead" is used to describe a system where the full cost of operation, including direct and indirect expenses (both internal and external), is budgeted by the audio visual organization. As far as the individual user of the service is concerned, it is "free"—no charge to the user for that service ever shows up on the user's own expense reports. The responsibility and accountability for how audio visual dollars are spent rest with the audio visual manager.

Of course, there is no such thing as a free lunch. Ultimately, the expense of an overhead operation is recouped from those business units within the organization that generate income. However, since AV is lumped together with all other support services, such as legal services, accounting and personnel, the units that do have to pay the ultimate bill seldom know just how much the audio visual activity contributes to expenses.

Direct Cost Allocation

The term "direct cost allocation" tells only half the story. In this system, internal overhead expenses are handled as full overhead and allocated as in a full overhead system. However, the users of audio visual services are charged directly for all external costs, such as talent, crew, outside production services, materials or anything else that generates a specific and identifiable invoice in connection with the project at hand.

For example, if a specific project requires videotape duplication as an outside expense, but all other services in connection with the production of the tape are handled internally, the client sees only the charge for the duplication. If the client is an income-producing business unit, then ultimately some part of the internal costs will get back to that client. However, the allocation will not be specifically identified with that project and may be more or less than the actual internal cost of the project.

There are two ways of handling the payment of the outside expenses. The AV department can pay the bill and then pass the charge through to the client as a credit against the AV budget, or it can simply pass the invoice directly on to the client for payment.

Full Cost Allocation

In the full cost allocation system, the expense of running the AV operation is charged or allocated back to the users of the service, usually on the basis of a percentage of the services used.

If the AV operation is responsible for the entire budget, both internal and external expenses, then both internal and external expenses are allocated on the same formula. In this kind of system, the allocation is usually based on a percentage of staff time. If only one client came to the AV unit for work during the year, then that one client would bear 100% of the cost of the operation. (If that were the case, however, most likely this would not be the budgeting system used.)

In some cases, the allocation is made to users for internal expenses only, while the direct external expenses may be handled in either of the ways described under direct cost allocation. In any case, the allocation of audio visual expenses usually shows up as an expense line on the client department's budget.

Another way of allocating these indirect expenses (called the "Massachusetts formula" and used by at least one regulated utility) is according to the number of employees in the various divisions and subsidiaries. Finally, AV expense may be allocated according to the percentage of income generated by the various business units. None of these, however, is a true allocation of audio visual expense, only a way of distributing overhead.

Chargeback System

The chargeback system is a method of allocating audio visual expenses to users of the services as in full cost allocation, except that a chargeback system makes a distinction among types of services. Rather than lump all expenses together and allocate them to users as a percentage of use, the chargeback system involves establishing a rate card for the various services provided. Users are then charged for the specific services, at the established rate and as they use them.

Most such systems require that each client establish its own audio visual budget on which it will draw throughout the year. The AV department also has a budget. As clients use the services, debits against their budgets are credited to the audio visual budget. As in the full cost allocation method, the objective is to parcel the entire AV budget out among all users, though in practice that rarely happens.

Unlike the full cost allocation system, a chargeback system makes a distinction among audio visual media. Because of the cost of equipment depreciation, maintenance and floor space, it is certainly more expensive to produce a videotape than it is to produce slides, even though the amount of staff time (the usual basis for allocation) may be the same. A hundred person-hours spent on television production is far more expensive than the same number of hours spent at a drawing board laying out slides.

Profit Center

The profit center system is similar to the chargeback system in that clients are charged on the basis of predetermined rates. The primary difference is that the AV department is itself a business unit. That is, it is charged with making a profit over and above simply recouping the annual operating budget. In this situation, the company is treating the AV unit as an investment, and the profit to be made is calculated on the size of the company investment in the unit (in the form of the annual budget and capital equipment). The AV department is then required to recapture the investment from the sale of services to clients, plus earn a percentage that is usually equal to what that investment would have earned if the money had been placed elsewhere.

Not surprisingly, most profit center organizations go outside their companies to generate additional revenue. Texas Instruments, Inc., for example, generates revenues by selling its technical training videotapes publicly, and the company requires that an annual profit be made by the AV department.

COST ALLOCATION SYSTEMS IN USE

The survey of the Audio Visual Management Association asked which cost allocation system members were using. As Table 2.1 indicates, the survey found that 19% of our respondents have what we call a full overhead budgeting system. The balance (81%) allocate or charge back all or part of their expenses in one form or another.

We also found that not all clients are treated equally: in 6% of the companies surveyed we find a combination of allocation systems. In most of these situations, the parent department is also the primary user of the AV services, establishing the budget for the activity and bearing most of the expenses as overhead. Clients ouside the parent department are then charged in one form or another for the services they use. Combination systems can also be used to handle budget overruns, where work is required which will force the AV department to exceed its annual budget. In this case, the client asking for the extra work is required to pay the cost, either directly or indirectly, in addition to any other allocations or chargebacks that may be made.

In Table 2.1, the combination systems are divided into two groups. The first allocates overhead to some clients and charges only direct expenses to other clients.

Table 2.1: Cost Allocation Systems In Use

Type of System	Percent Using
Full overhead	19
Direct cost allocation	30
Full cost allocation	19
Chargeback system	23
Profit center	3
Combination systems	
Allocation and overhead	3
Allocation and chargeback	3

Source: AVMA survey

The second allocates expenses to some clients and makes rate card charges for other clients.

Chargeback and Profit Center Systems

The prevalence of chargeback and profit center systems, totaling 26% combined, is worth noting. Further moves in this direction are likely. In fact, 24% of our respondents indicated that there had been a change in their budgeting and allocation systems in the past few years, and all changes were in the direction of cost allocation and chargeback. In addition, 4% noted that they expected future changes in their budgeting systems, and all were going either to a chargeback or a profit center basis directly from full overhead or direct cost allocation systems. These changes are summarized in Table 2.2.

Table 2.2: Changes in Cost Allocation Systems

Type of System	Previous	Current (Dec. 31, 1983)	Future
Full overhead	34%	19%	17%
Direct cost allocation	28%	30%	28%
Full cost allocation	18%	19%	19%
Chargeback system	12%	23%	25%
Profit center	1.5%	3%	5%
Combination systems			
Allocation and overhead	4%	3%	3%
Allocation and chargeback	1.5%	3%	3%

Source: AVMA survey

While only 66% of the clients for audio visual services were being charged for some or all of those services some years ago, within the next few years 83% will find themselves footing at least part of the bill for the work they request. This definitely creates a different attitude toward the audio visual media and a different atmosphere in dealing with the in-house audio visual department.

AV Budgets and Other Department Budgets

Are corporate AV departments treated differently from the rest of the organization when it comes to budgeting systems? The answer is yes, but not as differently as I might have expected. Only 40% of the AV departments surveyed use a budgeting system that is common throughout the rest of the corporate organization. Another 46% have budgeting systems that are similar to those in certain other departments, usually other service operations with similar functions within the parent organization, such as advertising or public relations.

However, the survey showed that 14% have totally unique budgeting systems, not found in any other department or function within the corporation. And, to my surprise, these unique systems are not restricted to chargeback and profit center systems, as I had presumed going into the survey. In fact, all of the profit center operations were similar to the budgeting structure of at least some other departments. Among chargeback systems, chargeback was common throughout the company in 20% of the cases and unique within the company in another 20% of the cases.

One explanation—and I know this has happened in at least two cases—is that the chargeback system has been developed specifically and uniquely for the AV department. However, once it has proven to be a worthwhile cost control measure, it is then applied to other departments. Sometime after this happens, even after chargeback has become common within the organization, the audio visual department then moves on to the next phase, that of profit center. It is not necessarily reassuring to the audio visual manager to realize that the department might be used as a guinea pig for cost control measures, but that could in fact be the case.

ADVANTAGES AND DISADVANTAGES OF COST ALLOCATION SYSTEMS

Naturally, there are advantages and disadvantages to operating under each of the budgeting systems. Some managers swear by the chargeback system, and others swear at it. That system generates the greatest amount of controversy and will be discussed in detail in Chapter 6.

In looking at each system, it is well to bear in mind that advantages and disadvantages will depend on a number of different factors. The accounting practices of the company, the type of company, and the stage of development and the client base of the AV department will go a long way toward defining which system will be "best." Naturally, a commercial production company must operate as a profit center. An AV department that serves a single client would be wasting its time to consider anything but the full overhead system. Most AV departments fall somewhere between those two extremes and may want to consider the pros and cons of the various systems as summarized in Figure 2.1.

How the AV manager perceives these advantages and disadvantages will depend on that individual and on the organization in which he or she must function. In many cases, moving into an allocation or a chargeback system may be the key to survival. In other cases, it can be a death knell. It is said that "clients never care what a project costs—until they get the bill." Most clients have no concept of the true cost of audio visual production. They see an ad for slides at $2 each and then become actually abusive when you have to charge them $25. They don't read the fine print in the ad for the $2 slides, and they believe that you are trying to put one over on them.

Some clients will complain loud and long about the prices charged by outside services, finding it hard to understand why a professional actor should charge $250 a day, a producer/director $500 a day or a photographer perhaps $1000 a day. It is interesting to note that chargeback systems are most common in manufacturing

Figure 2.1: Advantages and Disadvantages of Cost Allocation Systems

Advantages	Disadvantages
Full Overhead	
Is easy to administer.	Requires manager to justify requests from third party clients.
Gives manager a great deal of discretion in how funds are spent.	May mean some clients don't get projects produced when budget funds are exhausted.
Makes it possible for potential clients to "get their feet wet" in AV media without having to justify expense.	Requires no cost justification procedures for projects; company resources may be wasted on frivolous activities.
Requires no record-keeping of staff.	Leaves AV organization open to capricious budget reductions due to lack of cost justification and accountability elements.
Is of greatest value in new or developing operations or in departments with a single client.	Is of least value in situations where the AV department is attached to a function that does not use the AV service.
	Does not give clients a basis for realizing the real cost of AV.
Direct Cost Allocation	
Takes some pressure off manager to justify all production costs.	Gives client a distorted view of the true cost of AV services.
Helps control unnecessary expenses by charging external costs direct to client.	Puts pressure on manager to have work performed internally so client will not have to bear the cost.
Provides rationale for moving some services inside where economically feasible.	Often requires some explanation to client as to the basis for outside bills.
Puts burden of greatest expense on clients who incur them, especially when poor planning or heavy workloads may require buying services on the outside that would normally be handled internally.	Requires advance budgeting of outside expenses for client approval.
Requires only minimal administrative work, even if billing is handled by the AV department and re-billed to the client.	May tend to generate complaints regarding external charges.
	Requires extensive record keeping.

(continued on next page)

Figure 2.1: Advantages and Disadvantages of Cost Allocation Systems (cont.)

Advantages	Disadvantages
Full Cost Allocation	
Relieves manager of much of the burden of justifying the AV budget.	Is difficult and time-consuming to administer.
Generally allocates the greatest cost to the largest users.	Requires staff to keep time records when that is the basis for allocation.
Provides some measure of accountability for users.	Does not normally provide a project-by-project cost control and budgeting procedure.
Provides some measure of cost justification for the manager.	Tends to bury AV expenses within the allocations user departments get from other service organizations.
	Does not give the manager control over increases or decreases in client budgets.
	May be difficult to estimate workloads.
Chargeback	
Provides project-by-project cost measurement and accountability.	Presents an administrative burden, requiring record keeping by staff, and probably additional clerical staff and/or computer facilities to administer and operate.
Shifts the responsibility for justifying budgets completely to the client organizations.	Can create ill will among clients who do not understand relative costs of AV services and believe they are being overcharged.
Provides tools to measure productivity and compare costs with outside or competitive services and media.	Gives clients the opportunity to "comparison shop" on the outside.
Aids in the corporation's perception of the AV department as a business function.	Takes a degree of creative control away from AV staff when clients begin to ask for the cheapest rather than the best way to accomplish the given objective.
	Requires formal estimating procedures.
Profit Center	
Offers the same advantages as the chargeback system.	Presents the same disadvantages as the chargeback system.
If successful, demonstrates that the AV function can contribute to the corporate bottom line.	Puts considerable pressure on the manager to control costs and remain (or become) profitable.

companies, probably because such organizations fully understand the costs associated with producing a product.

CONCLUSION

When asked what allocation system they would prefer to work with, 34% of the managers in the AVMA survey expressed an interest in working under a different system then they now had. Assuming that the full overhead system is perceived as being the least restrictive system and the profit center as the most restrictive, we find that 55% would, not surprisingly, opt for less restrictive system. But, it is worth noting that the other 45% would like to operate under a more restrictive system. Fully two-thirds of that 45% would like to have a full chargeback system. There was even one entrepreneurial soul who stated he would prefer to operate under a profit center system. Table 2.3 summarizes these survey results.

Table 2.3: Budgeting System Preferences

Type of System	Now Using	Overall Preference	Selected by Those Preferring a Different System
Full overhead	19%	20%	22%
Direct cost allocation	30%	34%	26%
Full cost allocation	19%	13%	9%
Chargeback system	23%	25%	31%
Profit center	3%	3%	4%
Combination systems			
Allocation and overhead	3%	3%	4%
Allocation and chargeback	3%	2%	4%

Note: Of the total respondents, 90% indicated a preference for some specific system or combination of systems; 23% of those indicated a preference for a different system.
Source: AVMA survey

Individual responses indicated that some managers wanted chargeback systems so they could control their clients better and cut out the "Mickey Mouse requests." Another manager suggested that creative decisions should be "...governed by professional audio visual reasoning, not financial "whimsy," although this same manager also concedes that fighting chargeback systems may be "a losing battle."

Whatever allocation system an AV manager uses, it is still necessary to establish an annual operating budget, and that is the topic of the next chapter.

If a budget can be exceeded,
it will be.
Murphy's Law of Budgeting

3 Establishing the Departmental Budget

The annual budget is based on the fiscal year of the company. For most organizations the fiscal year is the same as the calendar year. However, the federal government's fiscal year runs October 1 through September 30, and state and local government fiscal years vary all over the lot.

Audio visual departments attached to corporate advertising departments may have a fiscal year that starts on August 1 or September 1, to correspond to the advertising budget. Or they may have production budgets beginning on September 1, while equipment and/or salary budgets begin on January 1. There are also cases where budgets are established for two-year rather than one-year periods. However, these variations do not really affect the process itself.

Even audio visual departments operating on a profit center basis are subject to the demands and requirements of establishing and living within annual budgets. In some cases, if the department makes a greater than projected profit, it may be able to use some or all of the excess profits to expand its operating budget, though this is not the practice at Texas Instruments.

BUDGET FORMULATION

Budgets can be formulated in one of two ways: as a percentage increase or decrease of the previous year's budget, or as a "zero based" budget.

Budgets Based on the Previous Year

If the budget is based on the previous year, corporate fiscal policy determines how much the increase or decrease will be and applies that as a constraint. Within the company the constraint may be applied across the board or selectively, with different departments being assigned different constraints. It is common to find the most restrictive constraints applied to service and support organizations, as opposed to the income-producing business units. For example, a manufacturing branch may be allowed an increase in annual operating expenses not to exceed 10%, while the audio visual department may be limited to 95% of the present year's expenses.

As we saw in Chapter 1, constraints are usually applied against actual expenses as opposed to budgeted expenses. The rule of thumb, then, is to spend to the limit of your budget. A department may budget $200,000 for audio visual expenses, but if it spends only $150,000, the next year's budget is going to be based on that actual expense, unless the department can prove good and valid reasons for not having spent the full sum. That is an uphill battle which can be tough to win.

Zero Based Budgets

In the case of the "zero based" budget, each year the department must treat expenses as if it were starting from scratch, even though staff and facilities are in place. Rather than simply being required to justify budget increases, the manager must justify the existence of the whole department. Zero based budgeting came into vogue in the late 1970s and was used by some federal agencies. I don't know of any corporate AV function that has to live with this system.

BUDGETING BASICS

Having established and gotten approval for the annual budget, is the audio visual manager then free to spend it at will? Usually not. Everything—the purchase of equipment, salary increases, new floor space rentals, or the purchase of outside services and materials—is subject to additional and continuing scrutiny at higher levels within the organization. The tap can be turned off at any time, either through the whim of an immediate superior or through a change in company policy or strategy. A strike, the loss of a major contract or some other catastrophe can have a major impact on that operating budget and can even result in elimination of the AV function altogether.

Budget Categories

While there are many variations on the budgeting process—far too many to detail in this book—we can look closely at the most prevalent. Let's begin with budget categories.

The annual budget is divided into five categories that fall into two general groups, internal expenses and external expenses. Internal expenses include salaries and overhead, while the external expenses are direct cash outlays for materials, services, equipment and construction. (See Figure 3.1.)

Figure 3.1: Budget Categories

Internal Expenses	**External Expenses**
Salaries and benefits	Materials and services
Overhead	Capital equipment
	Capital construction

Appendix C at the end of this book lists various expenses that fall into the non-capital categories. Salaries (with benefits) and overhead encompass all expense items of an internal nature (see Appendix C, Section A). The materials and services category can be broken into internal direct expenses (Appendix C, Section B) and external direct expenses (Appendix C, Section C). Equipment and construction are separate external expense budgets and will be discussed in Chapter 9.

Salaries and Benefits

In many companies, the salary and benefits budget is treated separately from the rest of the internal overhead expenses. The salary budget may also be subject to a different constraint than the rest of the budget. In many cases, the personnel department decides on the annual salary increase formula, so that the salary budget is a "dictated" budget.

This can cause conflict, however. For example, if the personnel department decides that annual increases averaging 7% will be granted, but corporate or department policy decrees that overall budgets will be subject to a 0% constraint, something has to give (or someone has to go).

Overhead

The overhead budget category includes a log list of expenses that are required for the continuing support of the audio visual department, its staff and facilities. Examples include rent, telephone, office equipment and supplies, employee awards, etc.

In most cases the company budgeting process will require that these be divided into individual budget line items, which must be treated separately. Sometimes, too, some of these line items may be subject to individual constraints—for example, travel or telephone expenses. Other categories, such as rent, cannot be controlled by the manager.

Materials and Services

The materials and services category covers all those items of expense that are related specifically to the work done by the department. It includes everything from videotape to hiring outside companies to produce complete projects. In some cases, project-related travel may be included in this category; more often than not it is lumped with other travel and treated as overhead. The materials and services category is subject to the greatest amount of fluctuation, since expenses are incurred only as required for specific projects. For this reason it is the most difficult category to project.

The Ratio of Internal to External Expenses

The ratio of internal expenses to external expenses can vary greatly. Some audio visual departments consist of only one or two people, who buy all services on the outside. In such cases, internal expenses may be less than $100,000, while external expenses can be well over $1,000,000. On the other hand, a large staff and a complete facility will incur a greater internal expense and a smaller external expense. Which is the most cost-effective and least expensive depends largely on the amount and type of work being done.

Location and media can also play a factor. State Farm Insurance built its major facility in Bloomington, IN, and hired a large professional staff; they felt they were too far from Chicago to be able to rely on that city's suppliers to meet their needs. On the other hand, Caterpillar Tractor Co., in Peoria, IL, produces films using outside suppliers exclusively and functions with a very small in-house staff of producers. In this case, however, the medium is film, and the films are shot on location around the world. State Farm's productions are videotaped, shot mostly in a home-office studio setting.

The Politics of Budgeting

Many managers (and not just audio visual managers) tend to take budget reductions and constraints personally. That reaction is not rational. Budget constraints and reductions are imposed by corporations in the interest of their own survival. Just as individuals cannot afford to buy everything and anything they want, so too corporations must weigh projected expenses against probable income and reduce those expenses, if necessary, to maintain profitability. It is the responsibility of the audio visual manager (as a member of company management) to work with, not against, the company financial staff toward that essential profitability. In other words, the manager must understand the "politics" of budgeting.

Understanding the Company

For that reason, the first step in audio visual budgeting is to get a full understanding of the company's budgeting process and financial situation. Take an accountant to lunch. Meet regularly with the company accounting and cost people. Don't be afraid to ask questions. It is essential for the AV manager to meet with his or her superior to get an understanding of company needs and financial objectives.

In our example of the Pimlico Corporation, the budget constraints were imposed like a "bolt out of the blue" at the last minute and after a lot of work had been put in developing what turned out to be a fruitless budget request. It would have to be a very poorly run company indeed in which budget constraints were suddenly applied in this manner.

The AV manager should maintain constant communication with company financial people so as to understand fully how the company as a whole is doing. Now, it is possible that our Pimlico AV manager's vice president wasn't keeping his own ear to the ground and so couldn't forewarn the manager of impending budget cuts. That would only make it all the more important for the AV manager to develop his or her own network to get this information independently. Even a review of corporate quarterly reports, couched as they are to show the corporation in its best light, will nonetheless give some clues as to what to expect in the next budgeting round.

In short, the AV manager cannot function in a vacuum. The penalty, when an excessively large budget is submitted, is to be labeled a spendthrift. The reaction from an executive can be, "What do you mean coming in here with a budget like that? Don't you know we're being wiped out in the marketplace?"

In the well-run corporation, constraints will be published far in advance, but not every company is well run. Furthermore, changing conditions can cause a revision of constraints even after they are published. The AV manager must be alert to the possibility of change and anticipate change when it occurs.

Some managers develop two budgets, one a fall-back budget to respond to reductions that may be required in a later round of budget negotiations. Of course, the fall-back budget is held in reserve until (and if) needed, but it certainly avoids the necessity of a lot of last-minute budget adjustments, as well as late nights and lost weekends.

Knowing Your Customers

The second essential step for AV managers is to know their customers. In our Pimlico example, the AV manager did survey the customers to find out what projects they would undertake in the next budget year, but he didn't himself ask the essential question: "Why?"

Yes, it is important to produce the projects that your clients and customers want, but it is also the responsibility of the manager to safeguard the resources of the corporation. The AV manager must be the first line of defense to assure that the projects produced are essential, have a cost benefit to the corporation and will contribute in some way to the objectives or bottom line of the business.

Far too often we fall into the trap of undertaking projects because they are exciting or will expand our creative horizons, without questioning the validity of the proposed projects. The budgeting stage is the time to ask those questions. The manager must always remember that he or she will be held responsible when a project is produced which raises the eyebrows of the corporate controller.

Knowing Your Own Department

The third preparatory step is to have full and complete knowledge of the present AV department budget and actual to-date expenses. Since budgets must be prepared several months before the current fiscal period is over, it is necessary to project actual expenses to the end of the period. Such a projection may bring the manager to the hair-raising conclusion that the budget will be exceeded. If so, it certainly helps to know it before it happens and someone else finds out first. And, since coming in under budget will cause an almost equal problem, it's important to know that in advance as well. In either case, it's necessary to find out why.

If the budget will be exceeded, which specific budget categories are involved? What projects or extra work caused that to happen? Furthermore, has this happened in prior years? A cost overrun in salary overtime for the past two years would seem to indicate that an additional staff member or more reliance on freelancers should be considered to reduce the overtime. Are travel costs excessive because of the types of projects that are being produced or because poor advance planning means that you are unable to take advantage of reduced air fares?

If expenses have been less than budgeted, the AV manager must know whether that is because some planned projects were not undertaken or whether it is, in fact, safe to reduce the budget for the next fiscal year accordingly. The manager must be prepared to explain exactly why certain expenses are under budget. If the manager is recommending that those budget categories should be kept the same or increased, a good explanation had better be available.

For example, in 1983, the cost of tape—both video and audio—was drastically reduced, so that many departments wound up under budget in this category. In this case, it may be safe to reduce the projected budget accordingly, but each such instance must be analyzed individually. The manager is going to need a clear crystal ball and a good financial understanding of the business to be able to develop a realistic and attainable annual budget.

THE BUDGETING PROCESS

Now let's do a fast rewind on our Pimlico AV manager to see how things might have been (or might be next year) if he had followed a better budgeting process.

Reviewing Proposed Expenditures

First, he began demonstrating his sense of fiscal responsibility to the vice president. He personally reviewed all planned and proposed expenditures, even though they were within the approved budget. Most were reasonable, and he himself understood completely why the expenditure was being made.

Expenditures that might raise questions, such as the need to hire a freelance photographer even though he had a photographer on staff, he discussed with the vice president in advance. He explained what he planned to do and why, and obtained agreement. Occasionally, the vice president didn't agree, but that was all right. Compromises are necessary and besides, by demonstrating an understanding of the rationale for not approving an expenditure, he gained the vice president's confidence.

Maintaining Good Communication

The manager and his vice president began discussing next year's budget well in advance of the deadline to submit it. They talked about the need to increase staff and decrease outside expenses. The AV manager proposed new management methods to control inventory, even talked about the possibility of installing a chargeback system and, in the process, got a good idea of what budget increases would be supported next year and in what areas there were likely to be problems.

At about the same time, the manager began making contact with the accounting department. He started by identifying the people who knew how the system worked. Initially, the contacts concerned small problems and questions. How is the equipment depreciation calculated? Is there a tax advantage to pre-paying bills before the end of the fiscal year? This was all by way of laying the groundwork for the big questions that would come up later.

Well in advance of the budget deadline, the manager began contacting his clients to ask them to submit their plans for the next year. He preceded his initial meeting with each client with a memo that asked them to begin thinking about their plans and to estimate the value to the company of doing each project they were proposing. After each client had had a week or two to think about this initial request, he made apppointments with each of them for a series of face-to-face meetings.

Estimating Costs and Priorities

At the meetings he asked his clients to describe in detail just what it was they wanted to do, the likelihood that the project would, in fact, be done, and how much they thought it was worth. He asked them to assign priorities to each project, assuming sufficient resources would not be available to do everything. He then took this information back and began estimating a cost for each project, both in direct external costs to the corporation and as a percentage of his own operating budget for internal expenses. He sent each client a listing of the projects, the cost and an overall schedule for all work to be done. Of course, he reviewed this information with his vice president before he sent it so as to have what might later prove to be invaluable support.

In sending the information back to the clients, he asked for their concurrence and a written agreement that the proposed list of projects and expenditures was essential, again asking them to reconfirm the priorities earlier established.

Adjusting Requests

As might be expected, a couple of the clients realized that the cost of some of their projects was much more than the return on the investment, so they dropped them. One client asked the manager to scale back a couple of projects to see if they could be done for less cost. One could and one couldn't, so the client agreed to drop the more expensive one.

As is not uncommon with marketing departments, however, that group still felt that they wanted to do everything and hang the expense (even though experience showed that they never did more than half of what they proposed). The AV manager went to his vice president for resolution of the conflict. Would the boss go along with the request, which would mean a budget increase, or would he discuss this directly with the head of the marketing department? As bosses are wont to do, he opted for the discussion, and, as often happens in cases like this, the marketing department backed down. After all, they didn't want to appear fiscally irresponsible either.

Monitoring Current Expenses

Of course, all through this process, and indeed throughout the year, our AV manager was monitoring his own budget, using some of the methods we will discuss in Chapter 7, so he knew he was going to be under budget for this year. Furthermore, he knew exactly why, and he knew that these would be nonrecurring expense reductions. He had already fully discussed this with his vice president.

The manager also knew by now that there would be a 5% constraint placed on next year's budget and that it would be over actual. Both the accounting department and the vice president had told him of this.

Preparing the Document

Having done all this groundwork, he was now able to produce the budget document quickly, with the full support of this clients and his boss (in fact, the clients' project proposals became a part of the budget document). He was also able to pinpoint just why he was under budget and why this could not be maintained in the next year.

To do all the work his clients felt was important, the AV manager would still be over that 5% cap. He therefore listed all the proposed projects and the anticipated cost of each, leaving it up to the budget review committee to decide which project or projects would be eliminated, or leaving it to his boss to argue the case.

He also prepared his fall-back budget, which in this case was based upon an increase of 5% over his present year actual assuming the budget review committee didn't buy his rationale. In this budget he was able to specifically identify the projects that would not be done if he couldn't get the additional funds.

CONCLUSION

Was his budget approved? If so, which budget? Answers to those questions are not important. What is important is the procedure. Far too often the AV manager is put in the untenable position of having to justify expenses on behalf of client departments and/or of having to explain to clients why the projects they want to do have been rejected. The method outlined above effectively takes the manager out of the line of fire. The battle is between the budget review committee (or its equivalent) and the client departments.

There is no question that doing this takes a lot of time. In fact, the manager may wonder whether there is time for the really important things like producing good programs. But that's the lot of a manager. Finding the time can be a real problem if the staff is very small—or if the manager is also the photographer, producer, writer, director and projectionist—but the question really is whether to take the time now or later. Furthermore, once such a system is in place, it will be a lot easier and smoother in future years.

The scenario just outlined will not work in every situation and may be overkill in some, especially if there are just one or two clients for AV work. In some organizations audio visual expenses are allocated according to the percentage of work done, but each client department must establish its own budget and get local approval. It is then only necessary for the AV manager to establish the cost of each proposed project to serve as a guide for the client. Even so, the manager should understand the cost benefit for each proposed project, in case a third party asks, "Why?"

The process also assumes that audio visual is growing and that clients are demanding more and more work each year. After the recent recession, not all clients could increase their audio visual budgets; in fact, many were looking for a reduction. As we said earlier, fully one-third of the AV managers in our survey were facing net budget decreases in 1984. In the next chapter we'll deal with some possible solutions to their problems.

It is useless to be a creative,
original thinker unless you can
also sell what you create.
David Ogilvy

4 Projecting and Generating Income

While the process of projecting expenses is important, the other side of the budget equation—projecting income—has at least equal importance. The corporate audio visual department's annual income projection is stated in the form of the budget dollars to be allocated to the activity during the year. On the surface it may appear that, once the budget is established, that is it; the income to cover the expenses of the operation is assured. However, this is not necessarily true.

Even in a full overhead cost allocation and budgeting system, it is essential that the audio visual manager continually monitor that budget. It is not uncommon to face unexpected reductions in mid-year. The manager must be forewarned about such a possibility and be prepared to react either with an offsetting decrease in expenses (which we'll discuss in more detail in Chapter 7) or by seeking alternate income. Such income can be generated from inside the corporation or from outside sources.

ALTERNATE INCOME AND AV GROWTH

Alternate sources of income not only offset budget reductions but also help the AV department grow. In the case of income generated from external sources, it can be a clear demonstration to company management that AV can make a positive contribution to the bottom line. Since audio visual is regarded as an overhead or "cost center" in most organizations, income generated can definitely result in a change of attitude about the value of the function.

However, the possibility of such a contribution to corporate profits should never be held out to company management either as a reason to create an AV department or to justify its expansion. The primary function of the audio visual department is to serve the needs of the parent company. Any income generated by the existing function is merely an offset to the expense of maintaining that function.

If the audio visual department is indeed to be run like a business, then it should also have the opportunity to operate like one. Just as the company as a whole can develop new products or services or seek new markets, so too can the audio visual department seek to expand its client base. Such an approach can have great appeal to the truly entrepreneurial audio visual manager.

Whether an increase in income will have an impact on a budget reduction depends on the type of cost allocation system in place. In a full overhead or direct cost allocation system, getting new clients or doing additional work for some existing clients would only increase, not decrease, the budget—unless it is possible to treat the new clients differently. For example, a department on a full overhead allocation system might find a new client who would accept either a chargeback or a full cost allocation for the services provided. This can be credited to the AV department's budget and will thus result in a direct decrease in that budget (at least to the extent that the work can be done with existing staff and facilities). Whether this can be done depends on company accounting practices and may also take a bit of wheeling and dealing on the part of the AV manager.

INTERNAL MARKETING TECHNIQUES

There are a number of techniques available to attract new internal clients.

Educational Techniques

Some techniques are educational in nature, with long-term results. They might uncover a client to meet the immediate crisis of a budget reduction—then again, they may not.

Newsletters

Informational newsletters can range from the fairly extensive and regular four-page publication put out by Ray Bonta at General Electric some years ago to simple one-page news briefs. The objective is to make current and potential clients aware of the scope and activities of the audio visual department. The publication should include news about current projects and staff and perhaps information about trends in the industry—such things as teleconferencing, computer graphics and other subjects that might kindle a spark of interest. The distribution should be broad, to all present and potential clients, perhaps to everyone at manager level and above, within a given corporation.

Public Shows

A public show might be an exhibit of photographs by company staff photographers or a continuous-play videotape shown in an employee lounge. The AV department could invite a broad-based group to a formal presentation, showing samples of its work and explaining how such things are produced. Of course, if the company has a regular videotaped news show for employees, this is a built-in promotional opportunity.

Targeted Presentations

The targeted presentation can be an invitation to selected present or potential customers to see a new production technique, such as interactive video, or a new piece of equipment that has just been installed.

Invitations to Meetings

Selected clients can be invited to attend local industry meetings covering subjects of interest. But be sure to pick the meetings, the subjects and the presenters very carefully. Relevant groups include the International Television Association (ITVA), the International Association of Business Communicators (IABC) and the Professional Photographers of America (PPA).

Selected Readings

The manager can send potential clients magazine articles or advertisements that might be of interest or that heighten their awareness of the potential of audio visual communications.

AV Handbook

The manager can produce and distribute an audio visual handbook. The handbook should explain what media are available, how to use them and, not incidentally, how the corporate audio visual department can help new clients to use the available media and services effectively.

Short-Term Approaches

While many of the above techniques are used by commercial production companies to generate long-term business, we should also take a leaf from their book for generating short-term or more immediate business.

Direct Solicitation

Account representatives maintain a list of potential customers. When they look ahead to a slow business period, they get on the phone and start calling. The corporate audio visual manager can do no less. In any company there are departments which may have a need for an AV program or perhaps meet that need through other sources, maybe even outside the company. They are potential clients.

It has often been said that "Nothing happens until somebody sells something." The audio visual manager, to succeed, must also make things happen by selling. The corporate manager has the advantage that the potential clients can be easily identified. First, the manager knows who uses AV and who doesn't. He or she also knows what departments do what kinds of things, as well as the condition of their budgets and their potential problems and needs. Of course, he or she has an inside track when it comes to talking to those potential customers directly.

Special Presentations and Demonstrations

Perhaps a presentation has just been completed using a technique or a subject that will be of interest to another department. The AV manager should arrange to meet with the manager of that department and show the program, talk about the possibilities and generally use the opportunity to feel out the customer.

Perhaps there is a department that either doesn't use AV at all or has its work done on the outside. Here the AV manager should arrange a meeting to demonstrate the capabilities of the in-house AV department and show samples of their work. Demonstration presentations should be available at all times for this purpose, even to the extent of producing a special program just for internal sales use.

There are some caveats. First, know your customer. Don't talk videotape if the customer's situation dictates slides as the appropriate medium. Don't talk high budget if the money isn't there. Don't talk communications if training is the objective. Don't oversell. Make your presentation and let it sink in. The results will not usually be immediate, but perhaps next week, next month or next year the client will call with a need.

It goes without saying that if a new client calls with such a need, the AV manager must be responsive. If the department is busy or the job doesn't seem big enough or important enough, it is still essential to follow through. If not, that customer will be lost forever.

METHODS OF TRACKING INCOME

Whether the AV manager faces a shortfall of income because some clients didn't come through with all the work they anticipated or because there is a general budget reduction, the need is the same—to market the audio visual product or services. But to know how much attention to pay to the marketing aspect of the job, it is also essential to keep tabs on that income.

Just as the manager must keep a weekly, monthly or quarterly log of expenses, so too a log of income must be maintained. Most expenses are regular and ongoing, particularly the internal overhead expenses such as salaries. But the income will vary with the workload. Therefore, the manager must know what work is anticipated, as well as what work has been done.

Full Overhead, Direct Cost and Full Cost Allocation Systems

The method used for tracking income will depend on the budgeting system in place. With full overhead or direct or full cost allocation systems, the manager will have determined specifically which projects each client will produce during the year. By maintaining a running list, the manager can know which of the planned projects has yet to be done. However, it is not enough just to assume that projects not yet started will in fact eventually get produced. Furthermore, the AV department would have a real problem if all the clients decided to save half their projects to produce in the fourth quarter—as many have a tendency to do.

Therefore, the manager will have to maintain regular contact with clients and continue to press for production schedules. Clients must be made aware of the conflicts that will occur if they wait too long to begin work. The manager can produce a quarterly schedule listing all uncompleted projects so that everyone can be aware of the work that has been projected by everyone else. Clients are just as interested in spending their allocated AV dollars as the audio visual department is in having them spent. Whatever the clients don't spend this year, they won't get to spend next year.

Chargeback and Profit Center Systems

In the case of a chargeback or profit center system, the manager should know the amount of each client's budget and be able to track expenditures against their budgets. If it appears that some clients are not spending enough of their budget dollars to cover the AV department's expenses by year end, the manager will have to find out why, as well as try to spread the workload—and the income—over a broader period of time. In Chapter 7, we'll discuss budget monitoring systems in more detail.

EXTERNAL MARKETING

In the early 1980s a curious phenomenon emerged. Concurrent with the recession and reduced corporate expenditures for audio visual projects, many corporate AV departments began to "go commercial," marketing their facilities and even their creative services outside their corporations. Some, such as American Express in New York, with its Blue Box Productions, have even gone to the extreme of establishing an audio visual subsidiary.

While the Audio Visual Management Association survey showed that only 6% of the corporations surveyed sold facilities time and/or production services outside their companies on a formal basis, another 32% did so informally. However, some of those that did sell or rent to outside customers commented that they didn't do it often due to busy internal schedules or that they wouldn't take on any big jobs because of the possibility of creating conflicts.

Obviously, this outside business is accepted, even solicited, as a way of generating additional income to offset internal budget reductions. A number of companies have built substantial facilities, with accompanying high overhead and operating costs. When work for internal clients is reduced due to budget cuts or the general recession, those operating and overhead costs don't go away. Thus, turning to outside sources of income is one way of making up the difference.

Guidelines

This approach causes philosophical as well as operating problems. From the philosophical standpoint, should the audio visual operation of a corporation be in direct competition with outside commercial production and facilities companies? Is it legal for them to do so? In many of the informal arrangements, such work may be in contravention of the company charter or of state law—not to mention the resentment it can engender among commercial production companies which depend on this business for their very survival. However, it may also be a matter of survival for the corporate audio visual department. If so, and if the company decides that taking such a course is in its best interests, we offer the following guidelines.

Fair Prices

Price the services fairly. As we will discuss in Chapter 6, many companies (and their AV managers) do not know the true cost of providing services and facilities. Before outside ventures are considered, those costs must be structured so as to recover the actual costs plus a reasonable profit. It is not sufficient simply to come up with a price for a specific service because it sounds good, and especially not because it will undercut the "competition," namely the outside commercial producer.

No Active Competition

Whatever the price structure established, don't actively market in competition with outside commercial houses in the immediate neighborhood. If your product or service is unique or is not available from other sources in the area, then fine, but don't attempt to steal business away from your fellow professionals. Bear in mind that the outside business you do is a supplement to your primary purpose, not a raison d'etre.

Supplemental Income Only

Don't hold out high hopes of significant income to your company. Again, that may or may not happen. The outside income should be considered "gravy," not a dependable income that can be counted on for basic budgeting purposes.

Equal Treatment

Give equal treatment to outside customers. In other words, if you're going to go into the business, do it professionally. The customer who buys services from you will not be back again if you can't meet his deadline because "we had to do this thing for the Chairman."

Good Business Practices

If you're going to do it, do it right. That is, prepare proper contracts and agreements. Handle your billing in an accurate and timely manner. Have someone available to serve as an account representative to your outside customers, and make your facilities clean, comfortable and professional looking when customers visit or work in your shop.

Selling Programming

There is one additional way of generating income, also from outside sources, that is somewhat less controversial—the sale of company-produced programming. It is truly amazing how many generic training programs are produced and purchased every year. And yet the market seems ever hungry for more. The reason is that a great many small companies have training needs but cannot afford to have their own AV production facilities or even to have audio visuals of any kind custom produced for them. Even larger companies have departments with small-scale training needs that do not warrant the production of a $20,000 videotape. However, they can certainly afford a few hundred dollars to buy commercial software for their minimal, and perhaps short-term, training requirements.

Conversely, many corporations can economically produce training programs for their own needs, which, with some minor modifications, can be made sufficiently generic to sell in the outside market. Obviously, this does not apply to proprietary programs where trade secrets might be jeopardized, but there are still many opportunities. The AV manager should be alert to these possibilities and recommend that outside markets be considered when appropriate.

The company does not even need to undertake the marketing itself. Usually, that can be arranged through commercial program distributors or trade associations, or even through manufacturers. For example, a program produced to train employees to operate a specific machine could be marketed to other purchasers through the machine's manufacturer. The manufacturer may even wish to buy the program and make it available free to future customers.

Programs in certain categories, such as health care, physical fitness, safety, home economics and so on can also be marketed to cable companies or manufacturers of products in those areas. The possibilities are boundless, and while the income will usually not be great, it can create a different attitude toward the AV department within the company. Now the department is actually making a contribution to the bottom line.

Joint Venture Production

An offshoot of this approach is what I call "joint venture" production. In a few cases I know of, training consultants have joined with the audio visual departments of major corporations to produce videotaped training programs. The consultants provide the expertise and content, while the AV departments handle the production. The resulting programs are then marketed, with the partners sharing the profits. In most cases, the programs are also made available to the training department of the AV department's parent company at reduced costs.

CONCLUSION

While it may appear on the surface that only expenses can be projected and controlled, it is possible to project and control income as well. There may not be the same degree of accuracy, at least at first, but with time both the amount of that income and the accuracy of the projecting will improve. It does require, however, a considerable time commitment from the manager. A little time spent planning in advance will reduce the amount of time required later on when the traditional sources of income begin to fail.

Both income and expenses are affected by the actual programs produced by the AV department. In the next chapter, we turn to the subject of production budgeting.

5 Production Budgeting

The audio visual manager (AVM) of the Pimlico Corporation was sitting in his office one day when in walked a potential client from the personnel department. The following conversation ensued.

CLIENT: Say, how much would it cost to do a sound-slide presentation, maybe 15 minutes long, to explain the company benefits plan to new employees?

AVM: Well, maybe you can tell me a little more about it. Do you need to explain the whole plan, or just get the employees to read the manual? In other words, do you want a technical or motivational presentation?

CLIENT: It should really get them to read the manual. Too much technical stuff for an AV, don't you think? Yeah, that's it—motivational. I'd been thinking: maybe getting testimonials from employees who'd had good experiences. I saw that done in one show and it looked pretty good...

We'll bypass the rest of the fact-finding; you get the picture. Let's assume that the AV manager did determine that sound-slide was the way to go and that the testimonial approach sounded workable. Ways could be found later to spice it up.

AVM: I think I've got the picture. Let me work out a budget for you, and I'll give you a call. If the price sounds right, I'll send you a written estimate for your approval, and we can go ahead with it right away.

CLIENT:*(Obviously in a hurry.)* Can't you give me some idea of the cost now? I mean, how complicated can it be? Besides, I'm not at all sure we want to do it yet, and I wouldn't want you to spend a lot of time estimating something we may not want.

AVM: *(Trying to be cagey.)* No, really, it doesn't take that much time, but just to be sure I give you the right number, it's worth it. I wouldn't want you to get a wrong impression.

CLIENT: Nah, don't worry. Just a ball park figure. I won't hold you to it.

AVM: *(Hesitatingly, but he's just been conned.)* OK, off the top of my head I'd say probably between $4000 and $5000, depending on how much location photography, the need for a professional narrator, music, that sort of thing.

CLIENT: $4000, huh? OK, I'll get back to you. *(He disappears out the door as the AVM is about to add a few more caveats.)*

(Two weeks later, same scene, same cast.)

CLIENT: *(Disturbed look on his face as he barges in the door with a sheet of paper in his hand.)* Hey, what's the meaning of this budget estimate?! $7500! You told me this thing would cost $4000.

AVM: *(Sheepishly and apologetically.)* Well, I guess I did mention $4000, but I said it depended on the amount of location photography and the need for a professional narrator. Besides, I didn't know you needed 25 copies of the show.

CLIENT: Oh, I'm sure I told you how many copies I'd need. Besides how much could duplicate sets of slides cost?

AVM: Actually, as I worked this out from your treatment, quite a lot. Looks like the copies alone will cost over $1500.

CLIENT: That still leaves this budget estimate $2000 over your original quote.

AVM: Then I didn't know just how many locations you actually wanted, and that you were going to want to go to three different factories and that you wanted computer graphics and...

(Slow fade to black.)

AVOIDING THE BALL PARK QUOTE

What has just happened is that the manager has lost the confidence of a potential client. It happened almost casually, but it occurs with great regularity. Furthermore, the scene is not restricted to the in-house audio visual department. Commercial producers too get calls to quote a "ball park" figure for producing anything from a dozen slides to a full-length film. Without benefit of script or storyboard, they have about five minutes to come up with an answer that will either result in no business or a continued professional relationship with that client.

The Pimlico AV manager did try one tack to stall off the client, but it takes guts to make it stick. Unfortunately, most audio visual managers are so hungry for business (as are most commercial producers) that they give in to that temptation and quote a "ball park" figure that will invariably come back to haunt them later.

There are a number of possible ways to deal with this not uncommon situation.

- Resist the temptation to quote that "ball park" figure. A client who has sufficient need to contact the audio visual department in the first place should have sufficient enthusiasm to wait for a formal price quotation.

- Turn the problem back to the client. Ask the client to provide more information about the project, including a complete statement of objectives, an audience profile and proposed scenarios—all in writing—after which an accurate cost quotation can be developed. At one time AT&T was asking clients to submit a four-page brief on each project, outlining the entire project in great detail. This included references to proposed cost savings as well, which gave the AT&T audio visual staff some idea of just how much the client might be willing to spend.

- Ask the client directly how much the project is worth. Very often a client has a cost figure in mind. If the client can be convinced to share that information, a production can often be designed to fit within the budget. If it's a totally unrealistic assumption, at least there is a basis for negotiation or for ending the discussion immediately.

- If a dollar amount quotation cannot be avoided, develop such a vague number that it cannot possibly be misconstrued. One approach might be to quote the cost of the most expensive presentation ever produced with an alternative cost so low as to be almost ridiculous. For example, "It's hard to put a dollar figure on that without more information, but I can tell you that the most expensive presentation we ever produced was $75,000. On the other hand, the cost of producing a simple slide is at least $25." Or, "Well, I can't really say. A photographer charges $400 a day, but we'd have to figure out how many days this might take." Commercial film producers often quote the industry average cost per minute of finished film. That quote is still used ($1000 per minute), and is just as meaningless today as it was 30 years ago—and just as valid in terms of serving the purpose.

- Offer to undertake the process one step at a time. For example, propose that a treatment be developed first, or a script, and that the cost for this phase will be X dollars. After that it will be possible to develop a truly valid cost estimate. If the client balks at this approach, obviously this will be a very low budget project. You should then attempt to find out just what the project *is* worth to the client.

- If there is a formal budgeting system, such as a budget worksheet or a computerized budgeting system, sit down with the client and start to develop the estimate right then and there. In most cases the client will realize the complexity of the budgeting process and go away, leaving the manager to develop the budget in peace.

- Every audio visual manager develops a technique to fit his or her own personality as well as the working environment. Unless there is an AV manager somewhere who has developed a technique to provide instant and accurate budget estimates, the foregoing list is the best I can offer. And if all else fails, quote high! Add 50% to the highest figure that comes to mind and throw that on the table. It is far, far better, and easier, to deal with a client who remarks later on how much lower your actual estimate is than to deal with one who has had an unhappy surprise.

THE BASIS OF PRODUCTION BUDGETING

The production budgeting process begins with a clearly defined and consistent information base. To attempt to provide a cost quotation for a teleconference when you've never done one is compounding the possibility for error. But with experience and over time, the cost for every project becomes a part of the production budgeting database. So, while I hate to harp on the need for record keeping, it is essential to develop a database of estimates given and actual costs for the completion of each project.

Factors Affecting Accuracy

The accuracy of specific project estimates depends on several factors.

Variety and Frequency of Projects

A major factor is the variety and frequency of projects undertaken by the audio visual department over a period of time. In the case of a teleconference, for example, even if one is produced every two years, cost estimates will be less accurate than for slide presentations, which are produced every week. Similarly, the fewer the types of media produced by the department, the more accurate the cost quotations. A department that produces only videotapes will generate more accurate estimates than one that produces a wide variety of media.

Ratio of Internal to External Charges

The more work that is performed outside, the more accurate the estimates will be. In effect, the responsibility to provide accurate budget estimates is shifted to the outside vendor or producer. If an outside producer is hired for a complete production, that producer is responsible for coming up with an accurate budget estimate. The audio visual manager may be asked to provide an initial estimate prior to hiring an outside producer for the project, but if cost is of paramount importance, the manager can shop for a producer who can meet the price, or the producer can be told to produce to the dollar estimate. Other outside costs such as tape or slide duplication can be estimated with great accuracy by simply quoting rate card figures.

The Cost Allocation System

The type of cost allocation/budgeting system will affect how estimates are made. Under a full overhead system, little or no project budgeting is needed. A direct cost allocation system leaves the manager with only the responsibility of providing cost estimates for outside charges. Under chargeback or profit center systems, the internal rate structure that has been established provides a rate card upon which budget estimates can be based. All the manager need do is estimate how many hours or the amount of materials to be used.

These systems also provide the database from which the AV manager can work. Since these systems are usually only in place where audio visual has been operating for some time, the manager has a chance of generating accurate budget estimates.

A full cost allocation system presents special problems that will be discussed later in this chapter.

The Complexity of the Project

Obviously, there is a much greater likelihood of giving an accurate estimate for the production of five slides than for the production of multi-image presentations for a national sales conference.

Knowledge of the Project and the Client

The fact-finding phase is the basis for accurate budget estimating. The more vague the client is about the stated objectives, audience and distribution of the end product, the less accurate the estimate is likely to be. For that reason the audio visual manager must be thorough and diligent in digging out every possible detail about the project before developing a budget estimate.

Knowing the client helps the manager to be forewarned about the possibility of expensive changes. Is this a client who changes his mind about objectives, has a hard time approving a script or is likely to want to make visual changes after the visuals have been produced? Or is this a client who is willing and available to meet with the producer on a regular basis to approve concepts, script drafts, storyboards, music, talent and sound tracks before final budgeting?

Developing the Database

The database itself is not hard to develop. As Figure 5.1 shows, many budget estimating forms contain spaces for entering both the estimated and actual costs for each line item. (Other estimating forms can be found in Part 2 of Appendix B.) By simply keeping these estimates on file, perhaps sorted by type of project, the manager will have a ready reference on the accuracy of previous estimates, as well as the basis for accurate estimates for future projects of a similar nature.

In some cases the database may contain actual costs. In others it may contain only the number of hours of staff time or hours or days of facilities usage. The amount and type of information will depend on the cost allocation system in place. However, even if the department is operating under a full overhead system, we recommend that time records be maintained. If and when (when might be more appropriate) the budgeting system changes, a database will exist which will enable the manager to convert to an allocation or full chargeback system with some degree of confidence that budgets and chargeback rates will be reasonably accurate. Furthermore, the necessary record keeping gets the staff accustomed to maintaining the logs that are essential in any allocation or chargeback system. A sample timesheet is shown in Figure 5.2. Even if the budgeting system never changes, the manager will have information that permits measuring productivity and facilities utilization, valuable data to have when budget time rolls around.

Full Cost Allocation Systems

The most difficult budgeting system under which to accurately predict project costs is the full cost allocation system based upon the percentage of staff time by project. It is difficult because neither the audio visual manager nor the client can know what the cost of each project is until all the costs have been allocated at the end of the accounting period.

Since the manager cannot know in advance just how many other projects might be undertaken or just how much time each will take, it is virtually impossible to provide an accurate estimate for the internal charges for a specific project. The unit cost per project will be relatively higher in a time period when fewer projects are undertaken, and vice versa. If the operating expenses of the audio visual department are $10,000 a month, the cost per project will be $2500 if four projects are produced and $1000 if 10 are produced.

The longer the accounting period, the more reasonable the unit costs will appear. However, to wait until the end of the year to be able to tell a client the cost for a specific project doesn't give the client control over his own expenses. Under this kind of system, the AV manager will have to have a better handle on current and projected workloads than under any other system; that is the only way to be able to give the client even a rough estimate of the internal costs for each specific project.

Figure 5.1: A Budget Estimating Form Showing Columns for Both Estimated and Actual Costs

```
AV Project Expense Report                        Date_____

AVC#_____ Title_____

Director_____

Client_____ Client Acct#_____

Department_____ Company_____
```

Expense Category	Estimated Cost	Actual Cost
Total Project Cost		
Staff Time		
Pre-Production	_____	_____
Production	_____	_____
Post Production	_____	_____
Artwork & Graphics	_____	_____
Photography	_____	_____
Supplies & Miscellaneous	_____	_____
Television – Studio A	_____	_____
– Studio B	_____	_____
– Location	_____	_____
Audio – Studio	_____	_____
Editing & Conforming	_____	_____
Special Expenses	_____	_____
Props & Sets	_____	_____
Animation	_____	_____
Rentals & Other Services	_____	_____
Talent	_____	_____
Duplication	_____	_____

```
        of the total actual cost $____will be charged to
        client's account numbers

        External Value Estimated (Purchase Price) Of This Project is $_____


                                        _____
                                        AV Control
```

Figure 5.2: A Sample Time Sheet

AUDIO VISUAL JOB PROCESS RECORD / AUDIO VISUAL WEEKLY TIME DISTRIBUTION																

NAME: _____   Page _____ Of _____   Week Ending: MO. DAY YR.

EXPENSE CD	LOCATION CD	JOB NUMBER	TYPE CD	DESCRIPTION	Rate	SUN.	MON.	TUE.	WED.	THUR.	FRI.	SAT.	TOTAL
			0 1										
			0 1										
			0 1										
			0 1										
			0 1										
			0 1										
			0 1										
			0 1										
			0 1										
			0 1										
			0 1										
			0 1										
			1	SUB TOTAL									
80210	0001		0 4	ILLNESS/EXCUSED									
80210	0002		0 4	VACATION/HOLIDAY									
80210	0003		0 4	A.V. MAINTENANCE									
80210	0004		0 4	MANAGEMENT TIME									
				GRAND TOTAL									
80210	0005		0 4	(LESS) OVERTIME									
				TOTAL REGULAR HOURS			7 8	7 8	7 8	7. 8	7 3		3 8 5

FORM 4961 NS (10/82)

Providing for Budget Contingencies

Every budget estimate must contain provision for changes, hedges or contingencies. The kind and degree will depend on the situation and the factors outlined above. Many producers simply state up front that the budget estimate is subject to plus or minus 10% or 15%. Specific expenses might be hedged; for example, the cost of the script will be X dollars, to include up to two drafts. Additional drafts or additional research will cost an additional, specified amount.

In a computerized budgeting system, a calculation can be built in to automatically add a contingency to various elements of the budget. This can be calculated differently for different elements. For example, television studio time could have a 20% contingency factor, while outside duplication could have one of 5%.

All budget estimates should contain a contingency in one form or another, the various Murphy's Laws being what they are. The contingency may be buried in the line items of the budget or may be clearly stated as a separate line item. There is a case to be made for separating it as a line item, in that the client might say, "If I meet my deadlines for approvals and am able to avoid changing my mind, I'll save some money." Unfortunately, what clients really say is, "How come you spent all my contingency? I didn't change my mind that often."

Put It in Writing

A verbal statement of the contingency is not sufficient. Clients have very short memories and certainly will never remember being told that the cost of this project was subject to plus or minus 10%. Always put the contingency in writing. It goes without saying that the budget estimate should also be in writing, even if this is not required. Verbal estimates, no matter how accurate, are subject to misunderstanding and misinterpretation and are simply not professional.

Provide for Revisions

A provision for budget estimate revisions and updating should also be made. While it may not be necessary to tell the client when the actual cost is expected to be less than anticipated (let him experience a happy surprise later—besides, you could be wrong), the client should always be notified of potential cost increases. It is, after all, the client's money, and the client should have the option of deciding whether a change he wants is worth the additional cost. If the client decides that it is, then a revised budget estimate should be submitted. While revised estimates should be submitted in advance of committing to the expense, this is not always possible. During production, the talent costs could double, for example, due to the need to shoot for one more day.

Production cost increases may not always be under control of the audio visual manager, either. Weather, illness or technical failures can cause increases in the cost of the project. This is the reason for building in contingencies in the first place. However, if the contingency is not sufficient to cover the added cost, the client should be notified as soon as possible. At this point, there is still the option of making some change that could reduce costs in other areas. Examples include cutting out a scene that has yet to be shot or duplicating fewer copies—even scrapping the entire project altogether.

In practice such things rarely happen. The client will usually accept the extra cost. Whether this is done willingly or with a lot of grumbling depends on how well the audio visual manager has handled the contingencies.

Sometimes a budget overrun in a certain category (even beyond the contingency provision) may be offset by lower than budgeted costs in some other category. For example, increased cost for travel might be more than offset by a projected lower cost for tape stock. Assuming the overall budget will not be exceeded, should such variances be reported? It depends. If the final expense recap is going back to the client line by line, then yes. What the client will see (and the only thing the client will see) are those line items which exceeded budget. The client feels he is entitled to any reductions, but overruns in any category cause concern. If the final expense report is simply a gross figure, then reporting line item variances is not necessary.

COMPONENTS OF THE PRODUCTION BUDGET

Production budgets are typically divided into two categories: above-the-line and below-the-line. In commercial production, above-the-line items generally include creative staff charges, overhead, producer's fee and other expenses that are normally fixed by the production company, including charges for owned facilities. Below-the-line items are for direct costs that are charged back to the client at cost or with a markup, such as travel, facilities rental, film and tape stock, lab services and duplication.

The same kind of division can be applied to the audio visual department. Above-the-line charges include all those elements that represent internal and overhead expenses, including salaries and benefits, rent, equipment depreciation and so on. Below-the-line charges include materials and services purchased outside for the specific project. These categories can also be referred to as internal and external costs.

The Three-Phase Budget

Both internal and external costs include a number of components that can be divided among the three project phases: pre-production, production and post-production. The pre-production phase includes all aspects of the project up to the point of actually beginning the physical work of producing the artwork or going into the studio or on location.

Pre-production costs include time charges, such as producer and director time spent in client and planning meetings, storyboard preparation, script development and set design time. Also included are some hard expenses such as travel for location scouting.

Production phase costs include time charges for the director or producer during the actual production process in the studio or on location, and for lighting and set designers during set build and light days. Other production costs include facility time and rental, talent, location travel and production materials. Assembly and programming of multi-image or slide shows as well as the final editing of videotapes or films (though editing is often referred to as post-production) are part of production. This phase encompasses everything required to turn the finished script into a master copy of the end product.

Post-production covers all expenses incurred after the finished product is approved—for example, duplication, distribution, and the time of the producer or director in assembling the final budget figures and completing the project work file.

As can be seen in Figures 5.3 and 5.4 and in additional budget estimate and project reporting forms contained in Appendix B (Part 3), some AV departments break budgets down by the three project phases. Others simply divide the costs into internal and external charges (see Figure 5.5). However, the use of the three-phase budgeting method does provide a greater degree of control for the audio visual manager. The method makes it possible to analyze the productivity of the organization in terms of pre-production versus production—for example, to see if it is taking too long to get projects into the works or if insufficient time spent on pre-production is driving up the cost and time required for the actual production.

Figure 5.3: A Budget Estimating Form Using the Three-Phase Budget

PROGRAM PRODUCTION- BUDGET

to be completed by <u>Program Director or Director, Special Projects following</u> approval of
PROGRAM PRODUCTION REQUEST Form.

Program:_____ Contact Person:_____

Series:_____ Phone:_____

Working Title:_____ Dept/Agency:_____

<u>Pre-Production</u>   Producer:_____hrs at $ ___/hr = _____
estimates   Other costs:_____
                               Subtotal:_____

<u>Production</u>
estimates   Crew: Camerapersons at $ ___/hr * ____hrs =_____
                Director   at $ ___/hr * ____hrs =_____
                TD   at $ ___/hr * ____hrs =_____
                Engineer(s) at $ ___/hr * ____hrs =_____
                Other   at $___/hr * ____hrs =_____
                                Subtotal:_____

<u>Post-Production</u>
estimates   Editing: Editor at $ ___/hr * ____hrs =_____
                    Director at $ ___/hr * ____hrs =_____
                Other Costs_____
                              Subtotal:_____

<u>Out-of-Pocket</u>   Talent   _____
                Graphics   at $ ____ per * ____ =_____
                Videotape: 1" at $ ___/hr * ____hrs =_____
                       3/4" at $ ___/hr * ____hrs =_____
                       1/2" at $ ___/hr * ____hrs =_____
                Transportation at 15¢/mi_____   _____
                Other expense_____ _____
                              Subtotal:_____

                TOTAL ESTIMATED EXPENSE:_____

Date:_____Signature:_____

Note: Return with PROGRAM PRODUCTION REQUEST to origination Dept./Agency.
Originating Dept./Agency Approval: _____Date:_____

NOTE: RETURN WITH PROGRAM PRODUCTION REQUEST to Asst. Station Manager.

Every project will have a somewhat different ratio of pre-production to production time. However, pre-production time is less expensive. It is certainly not worth rushing through pre-production if it is going to result in higher production costs later on.

Checklists

The accuracy and completeness of the production budget can be enhanced by a

Figure 5.4: A Budget Reporting Form Using the Three-Phase Budget

```
                    ALLOCATED (estimated) EXPENDITURES          ACTUAL EXPENDITURES

PRODUCER'S     Preproduction:
PRODUCTION
RECORD           Producer           ____mandays  $_____      ____mandays  $____
                 Content Expert(s)  ____mandays  $_____      ____mandays  $____
                 Typist             ____mandays  $_____      ____mandays  $____
                 Director           ____mandays  $_____      ____mandays  $____
                 Graphics Artist    ____mandays  $_____      ____mandays  $____
                 Remote Crew        ____mandays  $_____      ____mandays  $____

                 TOTAL              ____mandays  $_____      ____mandays  $____

               Production:

                 Producer           ____mandays  $_____      ____mandays  $____
                 Director           ____mandays  $_____      ____mandays  $____
                 Content Expert(s)  ____mandays  $_____      ____mandays  $____
                 Studio Crew        ____mandays  $_____      ____mandays  $____
                 Studio Overhead    ____mandays  $_____      ____mandays  $____
                 Editing Overhead   ____mandays  $_____      ____mandays  $____

                 TOTAL              ____mandays  $_____      ____mandays  $____

               Post Production:

                 *Validation        ____mandays  $_____      ____mandays  $____
                 Program Review     ____mandays  $_____      ____mandays  $____

                 GRAND TOTAL
                                    ____mandays  $_____      ____mandays  $____

FACILITY       Studio utilized:_____
               Studio dates scheduled from_____to_____editing facility from_____to_____
               Studio dates utilized from _____to_____editing facility from_____to_____

POST           Validation (attached) completed by_____on (date)_____
PRODUCTION     Special Distribution requirements:

               Recommended handout and reference materials and sources as well as pre-
                  requisite viewing:

               Program Review completed by producer on (date)_____
               Actual length of completed program:_____minutes   1" master    2" master

               Approved for distribution by:_____date_____
                                            _____date_____

NOTES:


*Does not include time allocated by sample audience as this should be part of their training
 assuming audience was properly selected.
```

production budget worksheet or checklist, as shown in Figure 5.6. Additional examples can be found in Appendix B, Part 2. Having a complete list of all the possible budget items, with the unit costs if applicable, helps to insure that nothing gets overlooked. (This is one of the reasons why computerized budgeting packages such as *Instabid*, discussed in Chapter 8, are so valuable.) To rely on memory to generate a production budget virtually assures that some element will be overlooked, especially if projects are complex or if the audio visual department produces a variety of media. It's like going to the supermarket without a list: you're bound to have to make another trip back. Only, in this case, your client may be very unhappy with you for making another trip back to his well.

Figure 5.5: A Budget Reporting Form Using an Internal/External Breakdown

```
AUDIO VISUAL SERVICES DIVISION                          DATE_____

AUDIOTAPE/VIDEOTAPE ESTIMATE WORKSHEET                  PROJ. #_____

ASSIGNMENT_____: PRELIMINARY_____: REVISED_____   MEDIA_____

TITLE_____   DIV. CODE_____

DIVISION_____CONTACT_____   TEL. #_____

PRODUCER_____TEL. #_____# ELEMENTS_____   # COPIES_____
**********************************************************************************************
INTERNAL CHARGES
                          No.          Unit          Item
Staff Time                Units        Cost          Cost

Producer (Hrs/Days)       ____      $ 60/425      _____
Technician (Hrs)          ____         35         _____
Extra Crew (Hrs)          ____         25         _____
Artist (Hrs)              ____         50         _____        TOTAL STAFF:
Photographer (Hrs/Days)   ____       50/300       _____        $_____

Facilities
Set Construction (Day)/Strike   ____    500/250      _____
Studio Production (Day/Night)   ____    1500/225     _____
Add'l Equip._____       ____    ____        _____
Add'l Equip._____       ____                _____
Location Video (Day/OT)         ____    900/150      _____
Video Editing (Day/Night)       ____    750/100      _____
Video Dubbing (Hr)              ____     75          _____
Video/Film Trans. (Hr)          ____    100          _____
Audio Studio (Day/Night)        ____    800/90       _____
Audio Dubbing/Dupe (Hr)         ____     60          _____
Prompter (Pages)                ____      3          _____
Audio Edit/Search (Hr)          ____     25          _____     TOTAL FACILITIES:
Equipment Rental                ____     ____        _____     $_____
Contingency................................................          $_____
                                TOTAL INTERNAL CHARGES               $_____
EXTERNAL CHARGES
Materials
Videotape 1/2 Hr (Master)       ____     75        $_____
Videotape 1 Hr (Master)         ____    110          _____
Videocassettes                  ____    ____         _____
Videocassettes                  ____    ____         _____
Audio Tape                      ____    ____         _____
Audio Tape                      ____    ____         _____
Audio Cassettes                 ____    ____         _____
Audio Cassettes                 ____    ____         _____
Film                            ____    ____         _____     TOTAL MATERIALS:
Other_____             ____    ____         _____     $_____

Outside Purchases and Expenses

Professional Talent_____Actors for_____days   $_____
Professional Services_____      _____
Equipment Rentals_____      _____
Messenger                                                 _____
Props/Sets/Furniture                                      _____
Building Services Charges                                 _____

Travel

Fares $_____X_____people X_____trips          $_____     TOTAL PURCHASES:
Per Diem: $125 for_____days X_____                 _____
Other_____
Publication:_____Copies at $_____
Administrative Services (10% or Max. $200/item)       _____     $_____

                                TOTAL EXTERNAL CHARGES               $_____

Comp. Date_____        TOTAL ESTIMATED COST                 $_____
```

Figure 5.6: A Production Budget Checklist

```
SUPPORT SERVICES/PERSONNEL/LOGISTICS BUDGET RECORD                    Page 2
    Item                        Supplier(s)              Proposed    Actual
 1. Writer(s)_____ $_____  $_____
 2. Graphics_____ _____  _____
 3. Scenery _____ _____  _____
 4. Props _____ _____  _____
 5. Tech Rentals_____ _____  _____
 6. Music _____ _____  _____
 7. Photography Stills_____ _____  _____
 8. Photography Motion_____ _____  _____
 9. Useage Fees_____ _____  _____
10. Shipping_____ _____  _____
11. Travel_____ _____  _____
12. Catering/Meals _____ _____  _____
13. Lodging_____ _____  _____
14. Research/Consultation_____ _____  _____
15. Casting_____ _____  _____
16. Narrator_____ _____  _____
17. Talent _____ _____  _____
18. Freelance _____ _____  _____
19. Other_____ _____  _____
20. Editing_____ _____  _____
21. Audio Mix_____ _____  _____
22. Contingency_____% Re-Edit, Over-Time, Equipment Failure, etc. ____  _____

23. TAPE SURCHARGE (For Production, Client & Library Tape Copies)  $200    $200
    PRODUCTION SUB-TOTAL ------------------------------------- $_____  $_____

24. Tape/Film Duplication & Distribution_____ _____  _____
25. Film Transfer_____ _____  _____
26. Promotion Writer/Photography _____ _____  _____
27. Other Post-Production_____ _____  _____

    POST-PRODUCTION SUB-TOTAL ---------------------------------- $_____  _____
                                        TOTAL BUDGET  $_____  _____
```

```
PLANNING CHECKLIST
a. Get AVP Budget/Treatment Approval_____    i. Is a Remote Survey Needed _____
b. Take Promotion Photos/Write Copy____       j. Is a Talent Briefing planned_____
c. Are Non-Disclosure Forms Required____      k. Has Engineering been briefed _____
d. Is Legal Review Required of Content____    l. Have you reviewed facilities bookings__
e. Is Feedback/Evaluation planned_____       m. Is Network Booked____TWX written___
f. Is a PR Module applicable_____            n. Have graphics spelling been checked___
g. Is Legal Review Required of Contract____   o. Is Prompter script in format_____
h. Review Compliance with Talent Union____    p. Get text for Cassette Labels_____
```

Note: This is the second page of a four-page project analysis worksheet. The entire worksheet is included in Appendix B as Figure B.9.

Budgeting by Available Funds

While we have generally been talking about developing a project estimate based on fact finding and objectives, it is also possible to develop an estimate based on available funds. If $5000 is available for a specific project, a $5000 project can be designed. However, in order to do this the manager will have to have an excellent database of costs on which to draw. It will also be necessary that the project be brought in on or under budget with no overruns allowed, so the $5000 must include a contingency provision. Furthermore, the AV department must have full control over the design and execution of the project.

PRESENTATION OF THE BUDGET

I haven't yet heard of any producer or audio visual manager who uses audio visual media to present a production budget—probably with good reason. While the budget is important, there's no sense overdoing it.

Too much explanation will cause problems. Most audio visual managers agree that clients are notorious nitpickers. Therefore, they make it a practice never to tell those clients any more than necessary about how the budget was established. (The same does not hold true for your boss, however.) To tell the client that the staff time budget includes 10 days of producer time is sure to evoke the question, "Why 10? Can't it be done in nine?"

Nevertheless, in spite of that caveat, some budget forms, such as the one shown in Figure 5.3, do give more detail to the client than is prudent. It depends on the client, but, in general, explaining a production budget to a client in detail inevitably makes that client a producer.

That practice holds true for commercial production companies too. The standardized bidding form for the Association of Independent Commercial Producers (AICP) includes a lot more detail than is necessary in most corporate situations, but even that form does not show the client just how the producer arrived at the dollar amounts quoted for director, producer, cameraman, etc.

The Estimating Worksheet

Therefore, the practice of commercial producers and some of the larger corporate audio visual departments is to use an estimating worksheet, from which a separate estimate form for the client is developed. The worksheet serves several purposes:

- It provides a checklist of all possible expenses to make sure that everything gets included in the final estimate.

- It provides a control and a database to permit comparing estimated versus actual expenses in a variety of specific categories.

- It provides a control for the project producer indicating, for example, how many days of producer time, how many slides, how many rolls of tape, etc., were expected to be used in the original budget plan. If the producer sees the possibility of exceeding the budget in any one of the specific categories, a red flag can be raised immediately, and the budget can be adjusted accordingly.

The Client Estimate

From the worksheet, the budget estimate for the client is derived. The actual budget estimate can be a simple one-line memo stating the cost for the entire project, or it may be divided into internal and external expenses (a two-line memo). Or it can be as complex as a four-page project report including everything from the costs by line item to a statement of objectives, projected cost savings and the production timetable.

The design of the actual form is up to the individual audio visual manager. But what every form must include is a line on which the client "signs off" (as in Figure 5.7). The form should be returned to the audio visual manager for the files—*before* production actually begins and before any of the client's money is spent. These are ideals and may not always happen. Still, some system should be set up to assure that signed forms are returned as soon as possible, even if the system is the manager personally keeping track of the estimates as they go out and following up to make sure the client signs off.

Presenting the Budget in Person

Some producers and audio visual managers prefer to present the budget estimate in person. This can be valuable, particularly with new clients. That way the client gets the opportunity to ask questions, and the audio visual manager gets to explain in a bit more detail what is involved in the production process, as well as how the budget was developed. (Beware of providing more information than is necessary, however.)

Meeting face-to-face can also help alleviate misunderstandings about the scope and intent of the project. Perhaps the client really didn't want to spend as much money as this project is estimated to cost. That can be ironed out, and the manager will get sufficient input to go back and generate a new budget estimate based on the revised project description.

BUDGET REPORTING

The project is completed, the bills are in, and another project for another client is in the works. If the department is operating on a chargeback or profit center basis, the budget reporting pretty much takes care of itself. A well-designed chargeback system will include a provision for analyzing budget versus actual costs by category, as well as for telling the client whether the project was under or over budget. If the project was over budget, and the client did not get a revised estimate, the manager may have to do some fast explaining, but the chargeback system provides the data to help do so.

If there is no automatic system, then the manager should provide the client with a summary report that shows the actual versus budget expenses. Once again, it is the client's money, even if the department is operating on a full overhead system. Providing financial feedback, even if not needed, can't help but enhance the image of the department.

Regardless of the system, the problem faced by most audio visual managers, and

Figure 5.7: A Production Budget Estimate Showing a Line for Client Approval

AUDIO VISUAL
PRODUCTION BUDGET ESTIMATE

AV AUDIO VISUAL SERVICES

☐ Assignment   Date _____
☐ Preliminary   Project # _____
☐ Revision   Media _____
Previous Estimate $_____   Div. Code _____

Title/Event _____

Chargeable Division _____ Contact _____

Project Contact _____ Tel. # _____ Div. _____

Producer _____ Tel. # _____ # Pieces/Elements _____ # Copies _____

Concept/Treatment _____   Paper Copies _____   Show Date _____

Script Due _____   Prod. Complete _____   To Dupl. _____

Copy Due _____   Preview _____   Dupl. Complete _____

Storyboard _____   Ship Date _____   To Client _____

BUDGET ESTIMATE

★ Staff Time . . . . $ _____
★ Facilities Time . . . . _____
★ Materials Charges . . . . _____
★ Building Services Charges . . . . _____
★ Talent & Casting Expenses . . . . _____
★ Professional Services: _____ . . . . _____
★ Licenses (Music/Stock Photo) . . . . _____
★ Equipment Rentals (External) . . . . _____
★ Facilities Rentals (External) . . . . _____
★ Props/Sets/Furniture . . . . _____
★ Teleprompter . . . . _____
★ Travel . . . . _____
★ Staging . . . . _____
★ Shipping . . . . _____
★ Duplication . . . . _____
★ Other: _____ . . . . _____
★ Contingency . . . . _____

INTERNAL CHARGES . . . . $ _____
EXTERNAL CHARGES . . . . $ _____
TOTAL PROJECT BUDGET ESTIMATE . . . . $ _____

Approved by _____ Title _____ Date _____

Please have this budget estimate approved and return to:

_____ , Audio Visual Services,

COMMENTS:

frequently voiced by them, is how to get the data they need to generate those final reports. The weakness is that expenses have to be reported. That's no problem with outside charges for which bills are generated, but how do you get your staff to report time and facilities usage? Frankly, that's a staff management problem, but it is generally conceded that getting good input requires that the forms designed to gather the input be easy to use. Forms *are* essential. To rely on a verbal reporting of time spent, usually rendered weeks after a project is completed, guarantees inaccuracy. And, while it is not necessary to have 100% accuracy, at least 80% is desirable.

Part 3 of Appendix B includes a number of different types of reporting forms. Some forms are designed to follow the job, as in a photo processing operation (see Figure 5.8), with entries made at each step of the process. Others are designed for use by individual members of the staff to make daily or weekly entries of the time or facilities used by project (see Figure 5.9).

Figure 5.8: A Budget Reporting Form Designed to Follow a Project

REQUISITION FOR PHOTOGRAPHIC SERVICES   24627
(Instructions on Reverse)

ORDERED FOR (Print Name)   EXTENSION   LOCATION   ACCOUNTING NO.   ACCOUNTING USE ONLY

APPROVAL SIGNATURE (See Instructions)   DATE   GRANT/SPF. NO.

SERVICES REQUESTED: _____

ON COMPLETION: ☐ MAIL   ☐ CALL   ☐ HOLD FOR PICK-UP

NEGATIVES			FOR PHOTOGRAPHY USE – DO NOT WRITE BELOW					
Pts. 75	Gross 77		Movie 88	Misc. Time 90	Prof. Time 91	Cash 92	Transp. 85	Port. 82
Mic. 78	Int. 83							
Copy 79	Misc. 89							
PRINTS DUE:								
First 76	Add'l. 84							
ENL./REDUCT. DUE:								
First 86	Add'l. 93							
>8x10 94	Add'l. 95							
SLIDES DUE:								
Mic. 286	Gross 289							
Gen. 280	Pt. 287							
COPY/DIAZO								
First 96	Add'l. 97							
DUPLICATES								
First 487	Add'l. 587		Job Completed _____					MC728/R881

3M uses a central computer system and optical card readers at each work station. Each individual involved in a project needs only to put a project card into the reader to generate input to the expense reporting system. Of course, if every member of the staff has a computer terminal at his or her work station, that can also be used to input to the budget reporting system. However, very few AV departments have that luxury. The need to generate paperwork for actual expense reporting will be with us for a long time.

Once the paperwork is generated, someone has to do something with it. Whether this means manually entering the individual reports on a central control form or entering the data into a computer, the reports will do little good unless they are collected and, in turn, used to produce a summary project report. Here again, a computer is a valuable asset, though if the size of the operation is small enough and the amount of information needed is not great, the job can be done by a clerk. To tell the truth, though, I know of more than one audio visual manager who has to do that part of the job.

CONCLUSION

It's two months since the opening scene of this chapter. The audio

Figure 5.9: A Budget Reporting Form Designed for Use by an Individual Staff Member

FILM/VIDEO
TIME SHEET

PROGRAM PRODUCTION

SPECIAL CODES

A 1 1ST SHIFT
   2 2ND SHIFT
B 0 REGULAR
   8 USER ALTERATION
   9 HOUSE ERROR
C 0 REGULAR TIME
   1 OVERTIME
   2 SUNDAY
   3 HOLIDAY

e.g. 181 = 1ST SHIFT,
USER ALTERATION
OVERTIME

COST CENTER	OPERATION	
70 STANDARD RATE	700	MEETING/CONSULTATION
	701	PREPRODUCTION PLANNING
	702	SCRIPT RESEARCH
	703	SCRIPT WRITING
	704	SCRIPT REVISION
	705	SCRIPT EDIT
	706	ART/ANIMATION PLANNING
	707	PRODUCTION PLANNING
	708	CAMERA OPERATOR
	709	PRODUCTION ASSISTANT
	710	EQUIPMENT SET-UP/STRIKE
	711	DIRECTING
	712	VENDOR COORDINATION
	713	TRAVEL TIME
	714	PROGRAM FILING/WRAP UP
	715	USER INCURRED WAIT TIME
71 FACILITY RATES AND LABOR	716	VIDEO EDITING - PRIMARY
	717	VIDEO EDITING - SECONDARY
	718	FILM EDITING
	719	SLIDE EDITING
	720	SLIDE PROGRAMMING
	721	SLIDE LIBRARY SEARCH
	722	FILM/VIDEO LIBRARY SEARCH
	723	MUSIC SELECTION
	724	AUDIO EDIT/MIX/DUB
	725	AUDIO ENGINEER W/EQUIP
	726	VIDEO ENGINEER W/EQUIP
72 VENDOR FACILITY - STANDARD RATE	727	VIDEO EDITING
	728	FILM EDITING
	729	SLIDE EDITING
	730	SLIDE PROGRAMMING
	731	MUSIC SELECTION
	732	AUDIO EDIT/MIX/DUB

NOTE: NON-CHARGEABLE TIME

1. LUNCHTIME: ENTER "L" ONLY IN JOB NUMBER

2. ENTER "N" IN JOB NUMBER
   ENTER SPECIAL CODE
   ENTER COST CENTER CODE
   ENTER CODE NUMBER FROM BELOW IN OPERATION NUMBER

900 VACATION
910 PERSONAL
920 ILLNESS
930 TRAINING & ADMINISTRATION
940 WAIT TIME
950 NON-CHARGEABLE CONSULTATION
960 MAINTENANCE
970 GENERAL & ADMINISTRATIVE
980 TEAM
990 SPECIAL PROJECTS
995 VENTURE PROJECTS

EMPLOYEE NUMBER   EMPLOYEE NAME

NORMAL WORKING HOURS   DATE   PAGE

STOP TIME / START TIME	JOB NUMBER	SPECIAL CODE A B C	COST CTR	OPER NUMBER	QUANTITY	COMMENTS		
						REQUESTOR	DEPARTMENT	ACTION

MT 636 REV 7/83

visual manager of the Pimlico Corporation is sitting in his office (almost hidden behind a mountain of expense reports) when the client from the personnel department sticks his head in the door.

CLIENT: Just wanted to let you know how effective that employee benefits program has been. The employees and the personnel reps have been real pleased with it.

AVM: I'm glad to hear that. It's always nice to get good feedback.

CLIENT: Yeah, and I was also pleasantly surprised by that final expense report. $9775, that's right on target for your estimate, considering those last minute changes we asked for. Glad you warned me about those extra costs—but it was well worth it. See ya.

(The client waves and disappears. The manager smiles and plows back into recapping his expense reports with renewed vigor.)

There's no such thing
as a free lunch.
Familiar adage

6 The Chargeback System

Like the mythical free lunch, there's no such thing as a free audio visual presentation, regardless of the kind of budgeting or cost allocation system in place. What the chargeback system and its offshoot, the profit center system, attempt to do is put a real value on the end product. If it isn't free, how much is it? The chargeback system also attempts to come up with an answer to the question, "Is it worth it?"

No single management topic engenders more heated discussion at AV industry conferences than the subject of chargeback. As was pointed out in Chapter 2, 28% of the respondents to the Audio Visual Management Association (AVMA) survey would prefer operating under a chargeback or profit center system, and 26% now do. The main attractions of chargeback systems from the operational standpoint are the degree of control they offer to the audio visual manager and the feeling they create of being in your own business.

The opposition feels that such systems take too much time, cost too much to administer and give the client too much input into the "creative" process through control of the purse. Both sides of the controversy are illustrated by a continuing stream of articles on the subject which appear with regularity in the trade publications.

JUSTIFICATIONS FOR THE CHARGEBACK SYSTEM

Regardless of how the audio visual manager may personally feel, the final decision as to whether or not to install a chargeback system is made by top management of the company. The decision usually represents an attempt to define the audio visual costs of the organization with more precision.

As the audio visual function grows within the organization, it comes under increasingly close scrutiny. Since quantitative judgments are very difficult to make simply by evaluating the end product, management needs some mechanism to measure the activity. Chargeback appears to be a way to accomplish this. Management's thinking goes as follows: "If we can't ourselves decide whether this service is worth the cost, let's put a price tag on it and see if the departments that use the service continue to use it if they have to pay for it directly."

Another question the chargeback system seeks to answer is whether the cost of the services provided would be less if purchased on the outside rather than being produced on the inside with owned equipment and facilities and a full-time staff. Emotional considerations aside, this is a valid management question that needs to be asked of every internal support service. Corporate computer, travel, printing, warehousing, art, publications and other support units not principally related to the company's main line of business are subject to the same scrutiny, or should be, or will be sooner or later.

In 1973, I made a presentation at an International Television Association (ITVA) conference on the subject of chargeback systems. One point made in the presentation was that even if a company didn't have a chargeback system in place and wasn't asking for cost information on audio visual production, it behooved the manager to have a system which told him or her what production was costing because, sooner or later, someone *was* going to ask.

Another point was that the production costs being quoted by AV managers operating outside chargeback systems were ridiculously low and misleading to the clients and the corporation as a whole. As I later wrote in the May 1974 issue of *Educational and Industrial Television*, "What the $200-a-tape, in-house producer better worry most about is that one day someone will find out what those tapes *really* have been costing." Regrettably we still hear audio visual managers quoting low production cost figures, although we may now hear them adding the caveat, "Of course, that doesn't include the internal costs."

Whether a chargeback system is now in place, whether there are intimations that management may require that one be installed, or whether the audio visual manager simply wants to hedge bets against one being installed in the future, this chapter will give some guidance as to the methodology involved in setting up and operating such a system.

THE OBJECTIVE OF THE CHARGEBACK SYSTEM

The basic data for the chargeback system are derived from the overall operating cost of the audio visual department, including all the internal costs of salaries, rent and other overhead, and the outside costs for the purchase of materials and services. In an allocation system, all of these costs can be lumped together and allocated back to the rest of the corporation, users and nonusers alike, on the basis of some formula or other.

In the chargeback system the objective is to allocate the total audio visual operating costs back only to the actual users of the services, in direct proportion to the degree of complexity of the services used—just as if those services were being purchased on the outside.

"Zeroing Out" Costs

Philosophically, management is looking to "zero out" the costs of the operation over the fiscal period. What management hopes is that the "income" derived from the use of all the audio visual services by various departments during the year will equal the total operating costs of the audio visual department. With a profit center system, management is also looking for a profit above and beyond the actual cost.

In either case, the charges for the services on a unit basis are designed to recoup those total expenses. The unit charges should be no greater than—and, one hopes, substantially less than—the cost of buying equivalent units of service outside the company. If the unit costs are higher, then the rational decision would seem to be to discontinue the internal service and buy those services from the outside. If the costs are lower and the organization is still able to "zero out" at year end, then the department is worth keeping and has proven itself to be "cost-effective."

Corporate Considerations

Practically speaking, however, there are some other considerations. First, as one audio visual manager pointed out, no commercial television production company in its right mind would build a television studio on the 26th floor of a high-rent office building. The fact is that the overhead incurred by a corporate audio visual department is going to be significantly higher than that of a commercial producer who puts his television studio in a loft building across town. Employee benefits are higher, too. Further, many companies have restrictions against hiring freelance or part-time people, and corporate salary structures are fixed and often non-negotiable.

Then there is the cost of services provided by other departments within the corporation, which may be allocated back to the audio visual department, including purchasing, accounting, law and personnel. The commercial producer would buy these services on an as-needed basis, perhaps handling some himself with existing staff. Certainly he would be free to negotiate with several suppliers for the best rates, but the in-house audio visual department is locked into using corporate services.

Some companies also have a general overhead allocation which assembles all of the costs that cannot be directly accounted for and spreads them back to all the departments within the company. This can include everything from the salary of the president of the company to the rent on the central mail room—again, expenses not incurred by a commercial producer.

Given all of these considerations, it becomes virtually impossible for the in-house audio visual department to have competitive rates and still "zero out" at year end. In fact, most audio visual managers now operating under a chargeback system will freely admit to not breaking even. If they do break even, it is usually because some overhead expenses are not included in their cost calculations.

In the course of the AVMA survey, one manager stated that he was going from a chargeback to a profit center format. When queried as to what that meant, he responded that it was based on management's decision now to include company overhead in his chargeback calculations. In other words, his idea of chargeback was simply to recover his direct expenses, while the inclusion of overhead automatically put him in the profit center class.

"Real" Money and Real Costs

There is another consideration that can impede management's objective to control audio visual costs. Chargebacks are often installed because top management perceives that audio visual services are being used "frivolously"; that projects being produced are not cost-effective, and that if the clients knew the real cost, they wouldn't waste company resources on projects that were not truly productive.

The fallacy is that under most chargeback systems the money is still not "real." Except in a very few instances where money does actually change hands (i.e., checks are written to transfer money from one department to another), the transactions take place through the centralized accounting system. Typically, this is a "paper" transaction wherein the audio visual budget of a user department is debited, while the audio visual department's budget is credited as the services are used.

So, while the user of the audio visual services may see reports of the actual dollar amounts involved, it has little direct or personal impact. The client is well aware that the cost of the audio visual operation is set and is not going to go away if that client doesn't make use of the services. The client is not going to save the company any money if the services aren't used. Furthermore, once the client has fought for and gained approval of a budget to buy those audio visual services, the client is certainly going to make every effort to spend all of that money, regardless of how cost-effective the resulting productions. This is especially true toward the end of the fiscal period, when the client enters the "use it or lose it" syndrome.

DEVELOPING A CHARGEBACK SYSTEM

Nevertheless, company management often decides that the installation of a chargeback system is the only way to identify and begin to control audio visual expenses. It now becomes incumbent on the audio visual manager to develop a workable system. It is highly recommended that the manager be as involved as possible in the process. Most chargeback systems designed by accountants will be unrealistic, unrelated to real-world situations, impossible to compare with outside purchase of similar services and a nightmare to administer.

The chargeback system develops in four phases: fact finding, rate setting, administrative system design and client communications.

Fact Finding

The first step is to determine what the actual costs of the operation are. Part of that process is to decide whether the chargeback should include all of the overhead cost allocations we discussed earlier. Whose salaries are to be included? In some systems, the department manager's salary is not part of the calculation; in others, it is. Management must agree to certain guidelines that establish the overall objective and define what is and what is not to be included.

After this determination, the place to start is with a line-by-line analysis of the annual budget. A chargeback system has three classes of project charges. The first is for internal expenses and includes all ongoing and general overhead, such as salaries, rent and depreciation.

The second category is for those charges that are incurred specifically and directly in connection with the projects undertaken during the year. This includes some expenses that might show up on an internal budget line, such as travel, but most are direct outside purchases of materials and services. These charges can be handled on a "pass through" basis. That is, they will be incurred by the audio visual department and then charged directly to the client either as used, in the case of bulk-purchased materials, or as incurred, as in the case of travel or outside facilities rental or duplication. It might even be desirable to get down to fine detail by including the basic cost of telephone service in general overhead but charging long-distance phone calls to the projects for which they were incurred.

The third category is for outside expenses that are incurred for the benefit of all clients or in order to provide certain kinds of services, such as equipment maintenance, computer timesharing and the purchase of music or photo libraries.

Determining Unit Costs

The purpose of fact finding is to develop the basis for a rate card. In the chargeback system, a unit cost basis will be developed for such things as producer time, hours or days of facilities usage or units of materials. In administering the system, each project will be charged for the number of units of each element used in producing that project: so many hours of producer time, so many days of studio time, so many units of videotape.

The next step in fact finding, then, is to allocate all of the identifiable charges to the specific functions. This is simplified if the audio visual department only produces in one medium. A full-service department that provides video, audio, film, photo and other media support functions will have a much more complex rate card.

Assuming the more complex situation, it is now necessary to determine how much of each of the individual types of expenses is attributable to which function. The annual cost of operating a television studio will include floor rent, equipment maintenance and depreciation, depreciation of test equipment, perhaps special charges for electricity and air conditioning, and the cost of salaries, benefits, desk space and overhead for the members of the staff required to operate and maintain that facility. It will even include the cost of sending an engineer to the National Association of Broadcasters convention every year to look at new equipment.

The cost of a photo lab includes similar expenses, plus the cost of chemicals, paper and other materials used to produce the end product. It is possible to include the cost of videotape in the television facility operation, but tape is generally a large expense item and individually identifiable by project, whereas photo chemicals and paper are not. Besides, we are working up to a different basis of charging for those materials, which will soon be evident.

Utilization Factors

The next step in the fact-finding process is to analyze the utilization factor in each element. How many hours or days will the television studio be used during an average year? Obviously it is not going to be used for the entire 1950 hours that would be available in a year (based on a 37½-hour work week). First of all there are holidays to consider, then there is maintenance time and finally there is just plain downtime.

Then, too, what is the demand? It is rare in a corporate setting to find enough demand to make use of the facility for 100% of the available time. Nonetheless, there are some facilities that are so busy that they operate on two shifts, weekends and holidays, and actually are in use 120% of the available time or more. The only way to determine the utilization factor is experience—this is where records of past use help.

The number of different utilization factors to be developed will depend on the organization, but to be consistent with commercial producers' practices, the following will apply.

- *Facilities:* Television and audio, divided into production time and editing (post-production) time, daily or hourly. Technical staff time may be included in the facilities rate or may be billed separately.

- *Staff time:* Hours or days of producer, director, writer, photographer or technician time. This is usually a separate item, since the time of these individuals is not necessarily tied directly into facilities usage. A writer will spend a great many more hours developing a script than will be spent in recording that script in the studio.

- *Units of production:* Where the end product can be measured in units, such as number of copies of a photo or videotape or slides, then the unit cost can be used to include materials, depreciation and maintenance of equipment used to generate the end product, and the time of staff involved in producing the product.

Just as the facilities rate will depend on the number of hours or days of use during the year, the staff rate will be based on the number of productive hours for each individual (or for all individuals as a group), and the units of production factor will be based on the number of such units expected to be generated during the year.

At one time, Deere and Co. had a chargeback rate structure based on minutes of finished videotape. This was simple to establish and simple to administer. All that was needed was to take the total annual operating expense and divide by the number of minutes of completed videotape produced during the year. The problem is that such a system does not meet one of the criteria of a chargeback system, which is to provide a meaningful comparison with the cost of providing similar services outside the company. No outside production company charges on that basis. Deere has since abandoned that system in favor of a more traditional method.

Comparison with Outside Costs

The final step in the fact-finding phase is to find out what the "competition" is doing. To do this, it is necessary to go on a rate card collecting binge. By getting rate cards from commercial producers for the various types of services, the manager will be able to determine average ouside costs for those services, as well as to analyze how such rates are structured. The internal chargeback rate structure should be based on the same kind of system so as to make comparisons possible.

Rate Setting

The assembled data may well contain anomalies. In one large corporation the audio visual manager complained that his accounting people had come up with a rate of $120 an hour for director time. Obviously, with a rate like that, there was no way he could show a lower cost for providing the service internally, since most freelance directors in his market were charging less than $400 a day. This is typical of what can happen when the rate setting is left to the accounting department and one reason why the audio visual manager must be involved in establishing the chargeback system every step of the way.

Rate Adjustments

This is also where the fun begins. The objective is not necessarily to break down each function to its individual elements. For example, in a television production facility, the pure cost analysis may show that the cost of editing is twice as high as comparable outside services, while the cost of studio work is only half as much. Since the two are interrelated, it is possible to adjust the rates to bring them in line with the outside rates while still obtaining meaningful data for the facility as a whole. In the photographic lab operation, the cost analysis may suggest that the costs are equal for printing color and black-and-white, but this is not the situation in commercial labs, so an upward adjustment in color print rates can be offset by a downward adjustment in black-and-white print charges.

By targeting the internal rates at some factor below the published commercial rate card rates (say 20%) and then using the utilization or unit of production factors to project income, it will be possible to know how and where internal rate adjustments can be made, and whether or not the operation can break even. Ultimately, it will also be desirable to generate a profit and loss statement for each function so as to know whether audio production—or television production or photography or whatever—is making money or losing it.

Rate setting can result in some highly unusual numbers, however. One corporate AV studio in New York City quotes a television studio rate of $150 a day. Clearly, this does not include all the cost elements necessary to operate that facility, since a comparable commercial studio would charge upwards of $3000 a day.

Achieving the Break-Even Point

For the audio visual manager, the bottom line is to break even (or show a profit). If all of the rate setting process shows that this can be done with a minimum of adjustment, well and good. If not, additional adjustments will be necessary. Remember, however, that the purpose of establishing the chargeback system is for company management to get meaningful information. In the event that the rate setting structure shows that the costs will be higher than outside commercial rates, this does not mean that the audio visual operation will immediately be closed down. Some time, usually a period of years, will be allowed for adjustments and to see how close the department can come to breaking even at what rates. Therefore, if the formal rate-setting structure indicates that the rates to be charged will have to be higher than those available outside, an arbitrary reduction can be made. At the end of the year the loss or profit can be analyzed and adjustments made if necessary.

One possible adjustment lies in consideration of the fact that the corporate audio visual department provides ongoing and available services to the company that would not be purchased outside. These range from giving advice to potential clients to developing and implementing long-range communications plans and even advising people on the kind of VCRs to buy for personal use. The convenience of having this expertise available is worth something—the question is, how much? In the case of a considerable and protracted series of losses, company management will have to decide if the convenience is worth the price.

Freedom to adjust rates should also be built into the system. If it turns out after a period of time that one function is making an exorbitant profit while another is losing money like it's going out of style, adjustments are in order. It should be possible to make them quickly and with a minimum of red tape.

Administrative System Design

When the decision to install a chargeback system is made, it must be pointed out to management that such a system will be costly to install and administer. A commercial producer can run his billing system with a part-time clerk and an accountant who visits once a month. Corporations don't work that way. A chargeback system for even the smallest of audio visual departments is going to be costly to administer because it has to interact with the existing company accounting department. Reports must be submitted in a form that is acceptable to and usable by that department.

The complexity of the system and how much it will cost to administer should be decided in pre-planning. Furthermore, if that cost is substantial, it too will have an impact on the rates to be charged, again inflating them. Chargeback systems for some of the larger AV departments cost more than $40,000 a year to administer. Who should bear that cost? The clients probably didn't want it in the first place, nor did the AV department. If at all possible, the audio visual manager should seek to have that part of the operating cost charged off as company overhead and not used as a factor in rate setting or in determining the profitability, or lack thereof, of the department.

Record Keeping

While there are a number of different ways that a chargeback system can be operated, basic to all of them is that input is needed from the people who do the work. In most cases this will take the form of some kind of input brief on which individual members of the staff record how many units of time or materials were expended in each project, as was discussed in Chapter 5. Many samples can be found in Part 3 of Appendix B.

The real problem is in getting the input in an accurate and timely manner. It probably has something to do with left-sided brain people versus right-sided brain people, but creative people in general are not sympathetic to paperwork. Furthermore, in the heady atmosphere of production, it is all too easy to forget exactly what was done for whom and when. Compounding the problem is the fact that the audio visual staff is usually not sympathetic to the chargeback system in the first place, viewing it as an encumbrance to creativity.

Keep in mind the saying "garbage in, garbage out." It is essential first that the staff understand fully the reason for the system and the consequences if it doesn't work. They might also be reminded that every commercial producer requires similar record keeping; therefore, this makes their operation closer to the "real world" of commercial production. The entire staff should have a vested interest in the profitability of the audio visual department. If that can include financial incentives, such as special salary increases and bonuses, so much the better.

Second, the forms and the system should be as simple to use as possible.

Designing forms that use "check-off" boxes or require only minimal entries will help assure accurate and timely input.

Finally, don't require a lot of detail. In the mid-1970s the federal government instituted a time-keeping policy in several departments. Staff members were required to note on a time sheet what they were doing every 15 minutes. A lot of the entries had to do with keeping time records, and the system was soon dropped.

Any time-keeping system that requires staff to account for every minute, or even every hour of every day, is doomed to failure. For that reason, it is recommended that reporting be done in daily rather than hourly units and that the staff not be required to account for the full number of hours every day, or even five days every week. By simple subtraction, it is easy to see how much nonproductive time is left over after deducting the chargeable time. The staff should not be required to write that down as well. The rate structure must take this into account.

Since the staff will also be required to charge time to individual projects, it goes without saying that a project numbering system should be set up, and all individuals should be aware of the numbers of the projects in progress at any one point in time. After a while, projects are likely to be referred to as much by number as by name.

Compiling Input

Once the input briefs have been received from the staff—and this should be done on a weekly rather than monthly basis—the information has to be collected, summarized and reported. How much time and sophistication is required to do this depends on the size of the operation and the form required by the accounting department. Ideally, there should at least be a microcomputer available to help in this process. Larger systems may involve data terminals tied to a company or outside timesharing mainframe computer. If this is the case, a computer program will have to be developed, and someone will have to be trained to operate it—either a present staff member or someone hired for that purpose.

The ultimate system is the one operated by 3M Company, which was mentioned in Chapter 5. In the labs and at various work stations are optical card readers. Each piece of work or project is identified by a card which can be inserted into the reader at each step of the process to record how much time is spent and the amount of materials used. This information is fed directly to the company's mainframe computers. Thus, the manager can instantly find out the status of any project as well as its cost as of that moment in time.

The system design should be rational. That is, a system like 3M's would be out of place in a five-person AV department. For that kind of a situation, a simple manual system would be adequate.

Assuring Timely Information

It is impossible here to explain and describe in detail the many ways in which input systems could be developed. What is essential, regardless of the system, is that the information must be timely. Remember that one function of the chargeback system is to offer a degree of project expense control to the audio visual manager. A system that doesn't tell the manager that a project is over budget until long after the project is completed is doing only half the job. The system should also include some provision for project estimating and for comparing estimated and actual costs at the conclusion of the job. More detail on this can be found in Chapter 5.

Whatever the ultimate form of the reporting system, whether a single handwritten sheet, the printout from a computer spreadsheet program or a detailed 40-page document, it ultimately finds its way to the client. This leads us to one of the most important aspects of chargeback systems.

Client Communications

Whatever the reason for the installation of the chargeback system or however it is to be administered, it is essential that the clients, who will suddenly perceive themselves as now having to pay for what they thought were free services, understand the system.

Sensitizing Clients

The first step, even before the process of establishing the system is begun, is to tell the clients what is about to happen. If possible, there should be a period of time during which the charges are reported "for information only," which means that while the clients are seeing the actual project cost reports, they don't actually have to pay them—yet. This first step in communications can be handled by memorandum or newsletter.

At the same time, a more subtle communications process should be implemented. This can range from casually telling clients what the cost for such-and-such a project would be if it were produced outside to sending them copies of outside rate cards for selected services. An immediate effort should begin to counter the potentially disruptive advertisements clients see that say they can buy slides for $2 or produce videotapes for $500. This is all part of the process of sensitizing the clients to the shock they are bound to have when they first see the audio visual department rate structure or get their first project cost statements.

Presenting the Rate Structure

When the rate structure is actually determined, a more formal communication process shold be implemented. Now that the rates have been set, meetings should be arranged with each client. The clients should be told once again why the system is being implemented and what it is to accomplish. They should be given a copy of the rate structure (no secrets at this stage). They should also be presented with some typical project cost comparisons. Using past projects as examples, they should be given copies of statements for what those projects would have cost had the chargeback system been in place at that time. These statements should also include a detailed estimate of what the cost for those same projects would have been had they been produced outside the company.

The shock of having to suddenly start paying for a formerly free service is going to be compounded by seeing what the cost of audio visual services really is. A client may believe that audio visual is the greatest thing since sliced bread, but at the same time have a real problem understanding why it costs what it does. A $50-an-hour charge for photographer time will result in a lot of snide comments about the high pay rate for photographers (though the client's salary is probably much higher), inferences that the photographer is actually pocketing that money and a failure to relate to the fact that the same client will willingly pay his plumber or automobile mechanic rates equally high or higher.

Continuing Communication

Since the installation of the system is being dictated by company management, the client will have little choice but to accept it. How that affects future relationships will depend a lot on continuing communication. The audio visual manager must take every opportunity, formal and informal, to continue that communications process. When rates have to be changed, the clients have a right to know why they are being increased, and, of course, rate reductions should be highly publicized.

Every audio visual manager I have interviewed is of the firm belief that communications is the key to the successful transition into a chargeback system.

CONCLUSION

Once the dust settles and the system is in place and operating, what happens to the audio visual department? There is a fear in the hearts of audio visual managers everywhere that having a chargeback system is going to cost them control over their product, result in clients hanging over their shoulders every step of the way or even put them out of business. The fact is I have not been able to turn up one case where these fears were justified.

On the contrary, many managers who have made the transition report that their clients reat them with more respect. They say that having a chargeback system has caused their clients to plan better, eliminated a lot of "Mickey Mouse" jobs and even resulted in more, not less, control.

An apocryphal story related to chargeback systems caused a lot of this fear in the mid-1970s. Standard Oil of Indiana in Chicago had a very large full-service audio visual department. When the company moved to its new headquarters building on the lakefront, it built a large television production facility on a high floor of that building. The department had formerly occupied quarters in a loft building some blocks away.

A couple of years after occupying that facility, the company decided to sell the studio to a commercial production company. The story got around that the chargeback system was responsible for closing down the audio visual department. Tom Richter, currently manager of that department, is quick to point out that he and 44 other full-time audio visual staff people are still there and operating with an annual budget that would make most commercial production companies green with envy.

What did happen was that the chargeback system pointed out quite clearly that it was economically unfeasible for the company to have its own high-rent television facilities in the middle of downtown Chicago. The company that did buy the studio went out of business soon thereafter, proving the point. With those high rents, they couldn't compete either.

In that case the chargeback system proved its value as a management tool. It identified a function that carried an unreasonably high price tag. Standard Oil could do far better renting facilities outside its own building. Furthermore, the trend to remote production resulted in less and less need for having a studio at all.

If the audio visual manager facing a chargeback system is filled with fear, it might be well to analyze that fear. Is it a fear of losing a job, of losing some of a carefully constructed empire of being "found out?" As a responsible manager, it is necessary to face those fears realistically and act ultimately in the best interests of the company. That's the best way to keep the job.

Beware of little expenses.
A small leak will sink a great ship.
Benjamin Franklin

7 Monitoring and Controlling Budgets

It is truly amazing how much time is spent in the process of developing an annual expense budget and in promoting or selling services to generate income, yet how little time is spent on the process of monitoring income and, even more important, expenses. The reason (or perhaps the rationalization) is that the day-to-day job of managing an audio visual department and of producing audio visual presentations is a demanding one. Very often the budget monitoring process is given a low priority simply because of a lack of time.

As was noted in Chapter 5 on production budgeting, it is relatively easy to develop a project budget but much more difficult to keep track of those expenses on a regular basis. So it is with the departmental budget, complicated both by the outside forces at work and by the complexity of the process.

The outside forces with which the manager must contend include the company accounting department, which must be relied upon for monthly or quarterly expense reports. Not only must those reports be timely, they must also be accurate. It is essential that these reports be checked in detail and compared to the manager's own records. When errors are found, they must be corrected immediately to avoid compounding the problem. It can take a lot of time to make an adjustment in an expense report, but that time must be spent to make sure that budgets will not be exceeded simply because of a clerical error.

Outside forces also include the audio visual staff itself, members of which may, in the course of day-to-day activities, commit the department to expenses which will cause budget overruns. Or the purchasing department may fail to negotiate a low bid on which you have based your budget, such as an annual contract for tape duplication.

KEEPING TRACK OF EXPENSES

The complexity of the process is illustrated by a quarterly audio visual internal expense budget report shown in Table 7.1. This is generated by the accounting department.

While such reports certainly keep the manager informed as to how he or she is doing in each internal expense category as well as in the area of external expenses (represented by the AV Materials line, code 274), the problem is knowing what to do with it once you've got it. Why are travel expenses over or under budget? Will that be corrected in future months due to more, or less, out-of-town production, and what impact will that have by year-end? To simply look at the bottom line and see that you are within budget is not sufficient. The next step is to take that information and project it to year-end.

The manager should keep a separate internally generated set of records against which these reports can be compared. Several forms shown in Chapter 5 and Appendix B are designed to do just that. A typical form permits entering expenses in various categories and comparing them to budgeted amounts in those categories on a month-by-month basis.

Controls on External Expenses

If the audio visual department receives and approves all invoices prior to processing by accounting, this provides a built-in checkpoint. As they are received, all invoices should be recorded and allocated to an account number, and the budget in that account should be debited at that point. It is better to confirm expenses at this stage rather than waiting either for confirmation from accounting that the bill has been paid or for a periodic report such as the one in Table 7.1, because these confirmations and reports are usually not generated until some time after the bill has been paid. They also may not provide the required level of detail. Furthermore, without an AV department record, the manager cannot know whether the amount actually paid and the report of those payments are in fact accurate.

In some companies, invoices for external expenses may be routed directly to accounting or purchasing for payment approval. The audio visual department only gets the information after the fact. If this is the case, the manager should seek some means of having those invoices routed through the audio visual department before payment processing. Most accounting and/or purchasing departments are more than happy to do this. After all, they are not anxious to have the audio visual manager jumping all over them for approving bills in the wrong amount, and they are just as glad to pass the buck to the audio visual department.

Such a system of prior checks should be installed no matter what the budgeting or cost allocation system in place. Operating under a direct cost allocation system, where invoices for outside services are paid directly by the client department, may seem to relieve the audio visual manager of the responsibility. One manager pointed out,

Table 7.1: Quarterly Expense Report

Division: Audio Visual	Code: 9996	Year:1984			Quarter: 2nd		Date: 20 Jul 84		
Expense Type	Code	Actual 2nd QTR	Budget 2nd QTR	Variance $	Variance %	Actual Y-T-D	Budget Y-T-D	Variance $	Variance %
Total salary	010	163,406	171,040	-7,634	-04.46	318,254	323,870	-5,616	-01.73
Outside help	020	0	500	-500	-100.00	0	500	-500	-100.00
Employee benefits	040	37,583	39,339	-1,756	-04.46	73,198	74,490	-1,292	-01.73
Salaries and benefits total		200,989	210,879	-9,860	-04.67	391,452	398,860	-7,408	-01.86
Space rent	050	51,228	51,225	3	00.01	102,456	102,450	6	00.01
Telephone and telegram	080	6,124	5,225	889	17.21	11,305	10,450	855	08.18
Travel expenses	090	3,847	11,400	-7,553	-66.25	12,352	22,800	-10,448	-45.82
Machine expenses	100	3,053	2,500	553	22.12	3,911	4,000	-89	-02.23
Data processing terminals	119	147	150	-3	-00.02	147	300	-153	-51.00
Data processing services	120	294	475	-181	-38.11	736	950	-214	-22.53
Outside timesharing	191	6,914	7,500	-586	-07.81	18,744	15,000	3,744	24.96
Bureau and association dues	228	558	400	158	39.50	933	800	133	16.63
Books and periodicals	232	254	300	-46	-15.33	302	600	-298	-49.67
F&E maintenance	254	78	80	-2	-02.50	255	160	95	59.38
Express/shipping	262	1,804	1,500	305	20.33	2,836	3,000	-164	-05.47
Professional fees	268	300	275	25	09.09	750	550	200	36.36
Employee awards	269	85	375	-290	-77.33	411	750	-339	-45.20
Other general expenses	270	1,447	1,100	337	30.63	2,329	2,200	129	05.86
Printing and stationery	272	0	250	-250	-100.00	0	500	-500	-100.00
Office supplies	273	1,022	725	297	40.97	1,715	1,450	265	18.28
AV materials/production	274	201,831	191,000	10,831	05.67	372,741	381,000	-8,259	-02.17
Building services	275	47,167	16,250	30,917	190.26	64,263	32,500	31,763	97.73
Total other expenses		326,153	290,730	35,423	12.18	598,186	579,460	16,726	02.89
F&E capital equipment	298	981	1,000	-19	-01.90	1,050	2,000	-950	-47.50
Division total expenses		528,123	502,609	25,514	05.08	988,688	980,320	8,368	00.85

however, that he prefers to have those invoices pass across his desk before they go to the client for payment. This avoids any unpleasant surprises or confrontations when the client complains about the size or type of a bill he is being asked to pay but didn't know he was going to get.

Controls on Internal Expenses

The same kind of controls and record keeping should be applied to internal expenses as well. Some things, like travel, are easy to control—simply keep a record of travel expense vouchers and airline ticket charges (the air fare can be taken off the ticket stub). The same applies to memberships, conferences and training, publications and other costs, such as computer software, that may be purchased by another department but are allocated back to the audio visual department.

Other costs are harder to keep track of. These include the direct charges by other company departments, such as accounting, legal or computer services, which may charge either directly or by allocation. It may not always be possible to know in advance just what such charges will be. However, before arranging for the services, try to get an estimate from the department in advance. Then watch the actual charges. If there is a significant, unexplained variance, ask why. It may be nothing more than a clerical error, or perhaps the charges result from using more services than are really necessary. Depending on the situation in any particular company, it may also be possible to renegotiate the charges to obtain a lower rate or a "volume discount."

Watch for mid-year rate changes, too. It is not unheard of to suddenly find an increased rent rate change in mid-year resulting from lease rate changes, property tax increases or other unforeseen situations.

Managing Controls

About now our AV manager from the Pimlico Corporation is starting to complain, "After all the time I have to spend developing my annual budget and promoting audio visual services in my company, not to mention my primary responsibility to oversee the production of cost-effective programs (and having to produce a few myself), now you're telling me I have to set up all these controls too?" As thoughts of working every night until 10 PM and spending weekends closeted in his den flit through his mind, we hasten to suggest that the manager does not necessarily have to do all this personally—after all, what is management anyway?

Mark Twain once said, "Never learn to do anything: If you don't learn, you'll always find someone else who'll do it for you." That may not be the most sage advice for the career-minded. But Billy Bowles, manager of audio visual for General Telephone in Tampa, told me once that he spent a lot of time writing scripts and directing productions, the things he felt he did well. When asked how he could do all that and manage a large department too, he answered that instead of hiring a producer to do all those fun things, he hired an administrator do do all the routine administration and bookkeeping—leaving him free to do the fun things.

So, instead of just taking an accountant to lunch, the audio visual manager might be well advised to add one to the staff. Of course, this would depend on the size of the operation and the attitude company management has toward that operation. If the operation is not large, a clerk can handle the record keeping on a part-time basis. If the department is too small to have its own clerical staff, perhaps a clerk or accountant can be borrowed on a part-time basis from another department.

The audio visual manager, however, is not relieved of learning enough about the accounting process to be able to communicate with the accountants (including the one on staff) or to establish and monitor the budget control systems. It is essential that the manager know what is needed, to insure that information is collected to meet that need and to be able to analyze the resulting data.

THE CONTINGENCY BUDGET

In Chapter 3, we mentioned the need for a "fall-back" budget in case the first one presented is not approved. It is just as important to have an ongoing fall-back or contingency budget plan available during the year. Getting an annual budget approved is no guarantee that the money will continue to be available during the year. It is not uncommon for changes in the economy, reorganizations, or company cash flow or profitability problems to result in a suddenly imposed reduction in the funds available for the audio visual operation.

Such budget reductions can affect the audio visual department directly, if that department is operating under a full or partial overhead system, or indirectly, in full allocation and chargeback systems. In the latter cases, general company budget reductions reduce the amount of money that clients have to spend on audio visual services and thus result in a loss of income to the audio visual department.

The audio visual manager can have two responses to such unanticipated reductions. One is to increase income. This subject is discussed in Chapter 4. But usually the effect of budget reductions is immediate, so—unless the audio visual department has a lot of clients standing by with additonal work to pick up the slack, or has a plan to generate immediate outside income—it will be necessary to institute expense reduction plans. Having a contingency plan ready eases the pain considerably.

In contingency planning the audio visual manager must know where the "fat" lies in the current budget, which specific expenses can be reduced and what the yield will be from each reduction. Except under the most strenuous and cost-conscious budgeting procedures, there will always be some excess in the final approved budget.

Reducing External Expenses

External expenses are the easiest to handle. Less work means lower external expenses, and the greater the ratio of external to internal expenses (the lower the proportion of overhead charges), the greater the impact. In other words, the smaller the AV department, the less the AV manager has to worry in the event budgets must be reduced, since less work will be required.

However, if the amount of work to be done is to remain the same and a budget reduction is dictated by the corporation, then reducing external expenses becomes much more problematic. We must assume that the audio visual manager has already negotiated reasonably good prices with outside suppliers. Now it becomes necessary to really sharpen the pencil on those outside expenses—to analyze them realistically to see if and where reductions can be made.

The following are a few possibilities:

- More efficient recycling of used audio and videotape, slide mounts or other reusable materials.

- Renegotiating purchases or negotiating bulk or annual contract purchases of supplies such as tape and film. Larger quantity purchases or the promise of large contract purchases during the year will usually result in lower unit prices.

- Combining purchases with a single vendor. For example, if two outside productions are to be done and scheduling permits, bidding both productions as a package will usually lower the unit cost. Buying all types of tape and/or film from a single vendor can have a similar impact.

- If you are in a high-cost metropolitan area, look for alternate vendors out of town. Lower costs available in other parts of the country may far outweigh the additional costs of shipping or travel.

- Can some services now provided outside be handled inside at less cost with additional facilities and staff? This has to be considered relative to cumulative budget impact. It is unlikely that the AV manager would be permitted to increase staff or facilities in the face of a general budget reduction. However, it may be the most economical path to take.

Reducing Internal Expenses

Reducing internal expenses can be more difficult. One approach is to eliminate all nonessential expenses such as organizational memberships, travel to AV industry conferences, subscriptions to publications or even excessive telephone use. The accounting department in one corporation simply removed every other telephone to reduce expenses.

Staff Reductions

When it comes to reducing overhead expenses, staff reductions are usually the first approach to come to mind and rightly so, since this can have the greatest impact. This can be done either through attrition (not replacing someone who resigns or retires) or by firing or laying off employees. The attrition route is the least painful, of course. However, corporate audio visual generally is a "young" business; few corporate audio visual people are ready for retirement.

Jobs in this business are also rather hard to come by, and there are a lot of applicants for every job that does open. As a result, resignations are not all that common either.

If the required budget reduction permits, and it would be difficult, if not impossible, to continue to function without a person altogether, it may be well to consider replacing an experienced employee who resigns or retires with an entry level employee. This has the disadvantage of requiring training for that employee (and perhaps temporarily impairing productivity). On the other hand, it not only reduces expenses but also provides advancement potential and thus motivation for that employee. In addition, there are savings in employee benefits. For example, replacing an employee who earns $20,000 a year with one who earns $15,000 yields a net savings of $6,500 when you include a 30% benefits calculation.

In any event, we do not recommend eliminating a position altogether except as an absolute last resort. It will be much more difficult and time-consuming to justify a staff increase later on when ultimately that position must be refilled.

While attrition is the easiest way to reduce salary expenses, it may be necessary to actually fire or lay off an employee to meet a mandated expense reduction. That is a difficult alternative, but it is a hard fact that any manager may face.

Other Reductions in Overhead

There are, however, other possible ways to reduce those overhead expenses. Some have been mentioned earlier. Here are a few more.

- If the budget includes a staff increase that has not yet been implemented, consider using freelancers or consultants to do the work at lower cost than hiring full-time employees.

- If there are charges for internal services from within the company such as for moving, cleaning, construction work, or accounting or legal services, try negotiating lower rates. If that can't be done, institute tighter controls over the use of those services to reduce their cost to the AV organization.

- While eliminating telephones might be an extreme, eliminating some types of phone services can help. Replacing multi-button phones with single lines, eliminating such things as call-forwarding, nonessential long-distance privileges or speaker-phones, or replacing phone company intercoms with inexpensive separate intercom systems can result in substantial savings.

- If reducing the amount of space used can reduce your rental charges, there are a number of possibilities. One approach is to clean house. Dispose of older files, equipment and other materials that take up valuable storage or office space. Consider moving stored equipment and materials to less expensive space. Perhaps even consider moving the entire AV operation to less expensive space, although the initial cost of constructing new facilities may far outweigh the annual rental savings for some years into the future. It might be possible to make work spaces smaller, consolidate offices and so on. The staff might complain, but when told that the alternative is to fire someone, the complaining should abate somewhat.

- Institute tighter controls over supplies. It is amazing how many pencils, how much paper, etc., just disappear over the course of a year. Don't do this to the detriment of the operation as a whole, however. Asking writers to steal paper napkins from the cafeteria to write scripts on is a bit extreme.

- Return unused or unneeded office equipment such as typewriters and dictating and adding machines.

- Institute tighter controls over travel. A company auditor was once heard to accuse an audio visual producer of going "first cabin" because that individual took cabs when public transportation could save money and submitted meal expense vouchers that were considerably in excess of the average.

Exactly what reductions can be made and what their impact will be depends on a lot of variables, but such opportunities do exist. It is surprising just how much impact a lot of these small reductions can have. However, it is important not to reduce the budget *below* the level required, since that would hurt the department when it came time to negotiate the next annual budget.

CONCLUSION

We cannot close this chapter without suggesting that perhaps the most effective and useful budget management tool is a personal computer with an electronic spreadsheet program. This is the purpose for which these programs were designed. They not only simplify continuous updating and maintenance and accurately handle the calculations, they also provide a ready means of doing "what if" projections for fall-back or contingency budgeting plans. If the audio visual manager cannot afford the luxury of an on-staff accountant, a personal computer is really the second best thing. The on-staff accountant would probably want one anyway. The use of computers and spreadsheet programs is the subject of Chapter 8.

The problem faced by all audio visual professionals who aspire to become "managers" is how to make the transition successfully from being a hands-on professional, perhaps perceived by company management as a "technician," to becoming a true manager. The management of budgets, income and expenses may be the most important step in that transition. When the AV manager at Pimlico is able to say to his vice president in July, "You should be aware that we expect to be 3% over our annual salary budget this year, primarily because of the extra overtime that had to be put in on that film the client kept changing his mind about. But we will be 5% under in our travel budget because we were able to do the personnel training project locally instead of going on location," then the manager begins to be perceived as a "Manager"—and that's what it's all about.

8 Computerized Financial Management

Although dedicated computers have been used for several years now in audio visual production for videotape editing, controlling multi-image shows and animation cameras, and producing slides, the use of computers for audio visual financial management is a fairly recent development.

Because the audio visual operation tends to be unique within the company and also relatively small, it has not generally been worthwhile to devote company computer resources to the development of the specific programs needed for AV financial management. Since AV professionals routinely work with both technical equipment and keyboards, they are often eager to take advantage of the potential of computers. However, all too often, corporate audio visual managers have been waiting literally years for computer programmer time to begin working on their specific departmental requirements. Nonetheless, the use of computers for AV financial management is developing, particularly since the advent of the micro-computer.

LEVELS OF COMPUTERS

There are three levels of computer systems. The largest is the mainframe system—the company's central data processing operation. This consists of central processing units and disk or tape drives having multiple-program and multiple-user capabilities. These systems can be accessed by computer terminals from anywhere in the world by telephone line. The systems are accessed on a timesharing basis, so that large numbers of users can be accommodated simultaneously using the same or different programs.

The next level is the minicomputer. These can be used on a stand-alone basis or in a small network. The number of users who can access the system simultanteously is usually less than 50.

The microcomputer is a stand-alone unit, usually for use by an individual running one program at a time. The Apples, IBM-PC and Commodore are all in this category. Microcomputers can also be used as terminals for minicomputers and even mainframes, with varying degrees of success and capability. We are also seeing developments in attempts to network microcomputers, despite such limitations as problems with file transfers and the fact that often only one remote terminal can access the computer at any one time.

The type of computer equipment available will determine the extent of the audio visual financial management system that can be installed. But as companies begin to encourage the use of more microcomputers, mere availability is becoming less of a problem.

What Computers Can Do

There are a great many financial management tasks that a computer can perform for the audio visual department, including:

- Preparing project budget estimates;

- Comparing estimates with actual expenses by project;

- Maintaining the department budget, including comparison of overall budget with expenses and balances to date;

- Generating project expense reports to clients;

- Keeping track of outside invoices payable and paid;

- Generating invoices for work performed for internal oι external clients;

- Preparing summary reports to management on income vs. expenses or on a wide variety of specific expense types.

In addition, there are a great many other things that a computer can offer the audio visual department. It is not the purpose of this book to delve into this area, but it's worth listing a few of them:

- Word processing for writing scripts and proposals, including direct conversion of written scripts to teleprompter formats;

- Creating slides and graphics for videotapes;

- Generating videotape edit lists;

- Programming multi-image shows—the multi-image show which opened in late 1983 at the South Street Seaport Museum in New York was programmed on an IBM-PC;

- Creating and maintaining videotape and slide libraries;

- Controlling materials inventories.

To use a cliche, the possibilities are limited only by imagination. Of course, a more realistic limitation is the availability of software to do the job.

Obtaining Software

There are three possible approaches to obtaining software. First, a custom program can be developed, but this requires skill in computer programming and an awful lot of time. A skilled programmer would take more than a year to develop a basic program package, and it would be commensurately expensive. For example, an estimate for creating an IBM-PC program just to generate vouchers to pay freelance production crew came to $2000, at the rate of $50 an hour. A somewhat more complex facilities scheduling program was estimated at $6000. It is just about as expensive to create computer programs as it is to produce audio visual presentations.

The second approach is to buy off-the-shelf general purpose software. There are literally thousands of programs available, from budget management and project management to word processing. The problem lies in the necessity of looking at a lot of different programs, considering whether they will meet the needs and then integrating them into the operation. The customizing and integrating of these programs takes quite a bit of time too, even though most off-the-shelf programs come with good instruction manuals. It took me the better part of a week to customize a spreadsheet program to do project budget estimating.

The third approach to obtaining software, which has only recently become possible, is to buy an integrated set of progams specifically designed for AV financial management. The rest of this chapter offers a closer look at off-the-shelf spreadsheet programs and the AV financial management programs that were available as of early 1984.

SPREADSHEET PROGRAMS

Off-the-shelf electronic spreadsheet programs provide a relatively inexpensive way to custom produce your production budget forms or manage departmental operating budgets. The electronic spreadsheets such as *VisiCalc, SuperCalc,* and *Lotus 1-2-3* allow the user to designate columns or rows and make unit or dollar entries. Then—and here's where the payoff comes—the computer does what it does best, which is make all the calculations automatically and correctly (assuming the user has set up the spreadsheet properly). Changes in numbers or units can be accomplished almost instantly, and the material can be printed out in a custom designed format.

The cost of such programs is generally under $600, often much less. Furthermore, spreadsheet programs are available for smaller computers, such as the Commodore 64, which would enable the AV manager to get into the computer age with minimal capital outlay.

The more sophisticated programs can also be integrated with database management programs, so that financial information can be exchanged with other programs such as word processing for producing reports.

One main advantage of these programs is that "what if" projections can be done. The manager can see what impact hiring an additional person would have on the budget or do a project estimate in different media or targeted to different budgets. Spreadsheet programs are particularly useful for estimating production budgets; I have done this with both *VisiCalc* and *Lotus 1-2-3*. The formats and the formulas used in making the calculations are basically the same for any spreadsheet.

The spreadsheet program develops a project worksheet and also a production budget estimate, which can be printed out and sent to the client for approval. The sample worksheet in Figure 8.1 is for a full-service in-house AV operation having graphics, photography, multi-image and audio as well as television production. It is designed for a chargeback system with pre-established rates for all types of staff functions, materials and facilities.

The numbers entered in the worksheet are automatically transferred to the production budget estimate form (another section of the same spreadsheet file) which is sent to the client (see Figure 8.2).

In more sophisticated spreadsheets like *Lotus 1-2-3*, some entries can be duplicated from the worksheet to the estimate form automatically, such as the project number, title, etc. Additional material would need to be entered if the user wished to include, as we did, project scheduling information. Such entries as the number of video and audio copies are also automatically calculated from the worksheet. Even the date of the estimate is generated automatically.

Each individual worksheet and estimate form can then be stored on a disk, should later revision be necessary, and of course printed out for use by the project producer and the client. The client may or may not see the actual worksheet. Often the problem with showing the client too much detail is questions such as, "What do you mean five producer days?! Four days should be plenty."

From personal experience I can categorically state that using the computer to generate project estimates not only saves time, both in doing the original estimate and in updating it if necessary, but also results in much more accurate estimates. In addition, the worksheet serves as a road map for the producer, helping to keep the project within budget.

Figure 8.1: A Representative Project Worksheet

Description	Increment	Rate	No.Units	Cost	Sub-Totals
Producer	Hour	50.00		0.00	
Technician	Hour	35.00		0.00	
Prod. Crew	Hour	25.00		0.00	
Photographer	Hour	50.00		0.00	
Graphic Desig	Hour	50.00		0.00	
			STAFF TIME		0.00
Set Build	Day	500.00		0.00	
Studio Prod.	Day	1200.00		0.00	
Character Gen	Day	150.00		0.00	
Graphics Cam.	Day	125.00		0.00	
Teleprompter	Day	185.00		0.00	
Video Edit	Hour	90.00		0.00	
Tape Dupe.	Hour	50.00		0.00	
Remote Video	Day	1000.00		0.00	
			TELEVISION FACILITIES.....		0.00
Audio Studio	Hour	40.00		0.00	
Music Search	Hour	20.00		0.00	
Audio Dupe	Hour	15.00		0.00	
			AUDIO FACILITIES..........		0.00
Multi-Image	Day	1500.00		0.00	
Conf. Audio	Day	800.00		0.00	
			CONFERENCE SYSTEMS........		0.00
Audio Tape	1/4"X10	40.00		0.00	
Audio Tape	1/4"X7	15.00		0.00	
AudioCassette	C-90	5.00		0.00	
AudioCassette	C-60	3.50		0.00	
AudioCassette	C-30	2.50		0.00	
1" Videotape	One Hour	125.00		0.00	
VideoCassette	C-10	35.00		0.00	
VideoCassette	C-20	35.00		0.00	
Videocassette	C-30	40.00		0.00	
Videocassette	C-40	45.00		0.00	
Videocassette	C-60	50.00		0.00	
			TAPE MATERIAL		0.00
Cat. 1 Slide	Each	25.00		0.00	
Cat. 2 Slide	Each	40.00		0.00	
Cat. 3 Slide	Each	60.00		0.00	
Dupe Slides	Each	1.10		0.00	
Photostat	Each	5.00		0.00	
Film 35MM	Frame	0.35		0.00	

(continued on next page)

Figure 8.1: A Representative Project Worksheet (cont.)

```
Film 120/35        Roll        14.00                    0.00
   Film 4X5        Each         3.50                    0.00
   Film 5X7        Each         8.00                    0.00
  Film 8X10        Each        12.50                    0.00
 Film 11X14        Each        18.30                    0.00
1st Prt 8X10       Each        16.50                    0.00
1st Pr 11X14       Each        30.00                    0.00

1st Pr 16X20       Each        45.00                    0.00
Glass Mounts       Each         1.10                    0.00
  Slide Tray       Each        10.50                    0.00


----------------------------------------------------------------------
                            PHOTO & GRAPHIC MATERIALS.        0.00
----------------------------------------------------------------------
   Talent Fees                                          0.00
  Casting Fees                                          0.00
Prof. Service                                           0.00
Licenses, Music & Photo                                 0.00
      Rentals                                           0.00
  Set & Light Design                                    0.00
   Props/Sets                                           0.00
       Travel                                           0.00
      Staging                                           0.00
     Security                                           0.00
Messenger & Shipping                                    0.00


----------------------------------------------------------------------
                            EXTERNAL PRODUCTION COSTS.        0.00
----------------------------------------------------------------------
Audio Cassette all lengths      2.00                    0.00
   U-Matic       10 Min.       16.13                    0.00
   U-Matic       20 Min.       19.50                    0.00
   U-Matic       30 Min.       22.10                    0.00
   U-Matic       40 Min.       25.60                    0.00
   U-Matic       60 Min.       29.80                    0.00
  Shipping   Mult. Dest.        6.00                    0.00


----------------------------------------------------------------------
                                       DUPLICATION           0.00
----------------------------------------------------------------------
                            TOTAL EXTERNAL CHARGES....        0.00
                            ------------------------------------------
    Producer:               TOTAL INTERNAL CHARGES....        0.00
                            ------------------------------------------
Project No.:                              CONTINGENCY         0.00
                            ------------------------------------------
                            TOTAL PRODUCTION ESTIMATE.        0.00
                            ------------------------------------------
```

Figure 8.2: A Production Budget Estimate Form

```
                      PRODUCTION BUDGET ESTIMATE
                      ~~~~~~~~~~~~~~~~~~~~~~~~~~~
        Media:                   Date:                    A/V #:

           TITLE:

        Div. Code:              Div. Name:

        Proj. Cont:             Phone #:              Division:
          Producer:             Phone #:           Compl. Date:
       No. Elements:      Video Copies:  0         Audio Copies:  0
       -----------------------------------------------------------------------
       SCHEDULE-
         Concept Due:           Script Due:           Storyboard:
         Prod. Begins:          Prod. Comp.:            Approval:
           Dupl. Comp:            Delivery:            Show Date:
       ~~~~~~~~~~~~~~~~~~~~~~~~~~~~~~~~~~~~~~~~~~~~~~~~~~~~~~~~~~~~~~~~~~~~~~~~~~~
                          Staff Time       0.00      Talent & Casting      0.00
                Television Facilities      0.00   Professional Services    0.00
                     Audio Facilities      0.00             Licenses       0.00
                    Conference Systems     0.00       External Rentals     0.00
                        Tape Material      0.00           Sets/Lights      0.00
            Photo & Graphic Material       0.00                Travel      0.00
                                       --------                Staging      0.00
                                                              Security      0.00
                   Total Internal          0.00    Shipping/Messenger      0.00
                                                           Duplication     0.00

                                                        Total External    --------
                                                        Total External     0.00
                                                        Total Internal     0.00
                                                           Contingency     0.00

                                        Total Budget Estimate....  --------
                                                                      0.00
       ~~~~~~~~~~~~~~~~~~~~~~~~~~~~~~~~~~~~~~~~~~~~~~~~~~~~~~~~~~~~~~~~~~~~~~~~~~~
        Approved by:_____Title:_____Date:_____

       Please have this budget estimate approved and return to Manager, AV Division.
       ~~~~~~~~~~~~~~~~~~~~~~~~~~~~~~~~~~~~~~~~~~~~~~~~~~~~~~~~~~~~~~~~~~~~~~~~~~~
           COMMENTS:


       ~~~~~~~~~~~~~~~~~~~~~~~~~~~~~~~~~~~~~~~~~~~~~~~~~~~~~~~~~~~~~~~~~~~~~~~~~~~
        ESTIMATE STATUS:     (Ass/Pre/Rev)    Prev. Est:

               PLEASE CONFIRM YOUR DIVISION CODE
```

AV FINANCIAL MANAGEMENT PROGRAMS

At this time, four programs specifically for AV financial management are available and are described below. For further information, contact the companies, whose addresses are included in Appendix A.

Instabid II

Instabid II was developed by Tom Fraser and Frank Slesinski of The Production Source. The Production Source is involved in a variety of videotape and film productions for both industrial and commercial clients. The company found itself needing either an additional person to handle the complex bidding and cost control aspects of its projects or a computer system to do the job.

In addition, The Production Source was in the business of hiring and paying production crews all over the country, a job requiring at least one more full-time person. In 1979 Fraser and Slesinski began work on the program that evolved into *Instabid II*, which they began marketing in 1983. The program was developed for the Altos minicomputer and is now also available for the IBM XT and Compaq Plus hard disk microcomputers.

In its Altos version, the system is a true multi-user, multi-tasking package. Up to three operators can work on the same or different programs at the same time using as many as six remote terminals. The program is licensed, and the licensee can obtain both the software and the Altos hardware from The Production Source. The program runs on Concurrent CP/M.

The IBM version was developed because so many companies already had IBM XTs; to be able to license only the software offered them a significant cost savings. However, there are limitations with the IBM. For one thing, *Instabid* recommends 20 megabytes (MB) of disk space to provide adequate file storage. The XT currently is available with only 10 MB. In addition, there are no multi-user capabilities with the IBM.

Instabid is a thoroughly comprehensive, well thought-out package. It contains a group of integrated software modules—some created from scratch, some off-the-shelf and some off-the-shelf with specific modifications to work with the system.

While the system was designed primarily for commercial production companies, ad agencies and the like, it is flexible enough to be easily adapted to a company environment for in-house work using internal rates. Packages have been licensed by AT&T and IBM for internal use.

Instabid is not inexpensive: as of January 1, 1984, costs ranged from $7500 for the basic IBM XT *Estimator* package without hardware up to just under $30,000 for the full-blown *Production Centre* system with the Altos hardware. (*Estimator* and *Production Centre* are discussed below.) Depending on the needs of the organization, production workload and available staff, it can be a very cost-effective solution, however. Use of the system will not only provide more accurate budget estimates and actual expense tracking but could also reduce the need for clerical or accounting personnel. It certainly is much less expensive than a custom-designed system.

The system is being updated and improved continually. Improvements are made available to all licensees at no additional cost within the first year. The program now also runs in MS-DOS languages on the Compaq portable hard disk system, which allows the producer to take his whole budgeting system into the field to keep tabs on remote shoot expenses on a continuing basis.

The Estimator

Instabid is available in two configurations. *The Estimator* is the basic package and provides a large number of schedules for calculating production costs. Built into it are such things as standard union crew, talent, benefits payments and even crew rules—for example, when overtime goes into effect, or special pay rates for weekends and holidays. Completing the worksheets generates a one-page budget estimate for the client which includes all project data and a schedule. Two different estimate forms, one designed by The Production Source and the other, the standard Association of Independent Commercial Producers (AICP) form, are included. The user can also design a custom form.

The real beauty of this program is that it is very easy to customize the individual elements or rates to your specifications. Any field or any calculation can be changed. The company will also do some limited customization for the licensee.

In addition to budgeting, the program permits entering both commitments and actual expenses on a line-by-line basis and provides automatic calculation of balances and percentages for each line item. It also permits recalling previous project estimates for direct comparison with current ones and using these previous estimates as the basis for developing new estimates.

The Production Centre

The expanded package, called *The Production Centre* has the *Instabid* program plus a general ledger, accounts payable and accounts receivable package customized by BizWiz Software. Thus, input to the estimated and actual expenses will produce a financial statement and a set of ledgers, generate invoices, pay bills and even print checks for vendors and payroll. Also included in the expanded package are the *Wordstar* word processing program and the *Microplan* electronic spreadsheet.

Looking ahead, Frank Slesinski sees the time not too far off when the computer could, for example, automatically call a crew person selected from the crew database file for a specific shoot date. If the person is available (confirmed by touch-tone phone), the computer will confirm the booking and the pay rate, generate the booking form, fill the crew roster position and, after the shoot, generate the payment check and post the expense to the general ledger.

Computer Aided Video

Like the programs from the Production Source, *Computer Aided Video* is a collection of programs, though not as extensive, at least in the financial areas. The collection was assembled by Comprehensive Video Supply Corp. of Montvale, NJ.

Individual programs include *Datafax*, which is an electronic filing system; *Edit Lister*, a videotape editing program which generates CMX-340-compatible punched paper tape; *PowerScript*, a word processing program specially adapted for generating film and television scripts; and the program of most interest to use here, *The Associate Producer*.

The Associate Producer is actually a package of four programs, designed by Lon

McQuillan. It is available in versions for the Apple and Apple-compatible computers, for CP/M-language computers like the Sony SMC-70, and in MS-DOS for IBM and IBM compatibles. Two of the programs, *Production Schedule* and *Program Rundown*, are production planning aids and are not of primary interest to us here.

Production Budget and *Budget Tracker* are an interrelated pair of programs which can budget above-the-line and below-the-line expense categories for individual projects. There are 134 pre-entered line items, 54 above-the-line and 80 below-the-line, divided into production and post-production categories. The pre-entered items can be changed, rearranged and generally customized to the user's needs. Comprehensive can also provide some specific customer tailoring.

Once the listing of expense categories has been set, *Production Budget* prompts for an entry for each item and permits entering either a flat fee or a number of days or hours, and then a rate per unit. The user can set up a standard rate table, which will then automatically be entered for each new job, thus eliminating the need to enter the unit rate each time. Like the other programs discussed here, the program provides an automatic checklist of production categories and allows making "what if" projections. It also provides for making short notes with each production category so that the name of the individual or the type of equipment can be specified right on the budget sheet.

The program will then print out the entire project budget with totals by category and sub-category (such as personnel or talent costs) and will also provide a summary sheet. However, the program does not produce a formatted budget estimate for client approval and sign-off. This would be a useful addition and is available in the other programs discussed here.

Budget Tracker works from the project estimate developed by *Production Budget*. It prompts for entries on expenses as they are incurred, by line item. This will produce a comparison of budget with actual and a summary report of the balance available (or the overrun) in each category. It cannot input commitments as compared to actual, but specific categories can be updated as bills come in. A substantial number of project budgets can be stored and manipulated, depending primarily on the capacity of the computer in use.

The program is quite easy to set up and run, and operates very fast, at least in CP/M configuration. Though it is not as sophisticated as the *Instabid* package it is also a lot less expensive, priced at only $550 for the MS-DOS and CP/M language versions, and $500 for the Apple. Comprehensive offers a demonstration disk for $10 to permit evaluating the program.

AICP Production Budgeting Software

The Association of Independent Commercial Producers has standardized a production budget form which is used by advertising agencies and is often the only bid estimate form accepted by them. *Instabid II* will generate an AICP standard form in addition to other forms. However, some other companies have developed specific AICP-compatible budget estimating software which is being used by many of the larger commercial producers. Two packages of which we are aware are one from Pacific Analog and one called *Bidform*, distributed by bt Systems, Ltd.

These programs run on the IBM PC and, in the case of *Bidform*, also on the Apple IIe. While not as sophisticated as *Instabid*, they are more complete than the Comprehensive programs. However, as in almost everything else, you get what you pay for. These programs generally sell for between $3000 and $5000 and might be worth looking into if you are planning to bid on much commercial production.

CONCLUSION

As we noted earlier, the use of the power of the computer to manage audio visual finances is really just in its infancy. However, more and more frequently we see magazine articles on the subject, and new software is under development constantly. This chapter has touched upon only a few of the possibilities, and a lot more sophistication and capability should become available in the near future, particularly as microcomputers become more common in business.

The near future holds many possibilities that will help the AV manager become more cost-effective and more efficient. Taking Frank Slesinski's vision of the future of *Instabid* a little further, I see the day when every AV professional will have a terminal at his or her desk to write scripts, produce project budgets, input to departmental budgets, schedule meetings and projects, access files of outside services and talent, make phone calls, send messages, book airline reservations, generate video edit lists and even do offline editing, program multi-image shows, select stock photos and music (with pictures and sound), look up letters and proposals and, I suppose, even make coffee if desired. Furthermore, the desk may be at the office, at home or in a hotel room while on location in Paris, Indiana, or Paris, France.

*If you're going to have to sweat
to get approval to spend money,
you might as well make sure
it is a lot of money.*
Anonymous

9 Capital Budgeting

The consuming topic of discussion at audio visual industry meetings seems to be equipment. How good it is, how to get it, how to upgrade it and, of increasing importance, how cost-effective it is. But discussing equipment with one's peers at an industry meeting is one thing. Discussing it with a corporate executive who has no idea what SMPTE time code is (and couldn't care less) and who can't see a difference in 10db signal-to-noise is quite another matter.

Like it or not, the audio visual business is equipment dependent. Sounds and pictures cannot be reproduced or displayed without electro-mechanical assistance. But therein lies the problem. For example, the Audio Visual Management Association survey revealed that 59% of the respondents report to creative departments, including public relations, advertising and so on. Historically, these departments have not had significant capital budgets. They deal in concepts and ideas and rely on outside services such as film production companies, advertising agencies or printers to bring those ideas to fruition. Other than relatively inexpensive items, such as typewriters and word processors, photostat machines and a few other odds and ends, those executives have not been required to seek capital budgets—until the advent of audio visual.

I also suspect that many of the executives who run such departments tend to look at the equipment-intensive operations, such as the printing plants, with some degree of disdain. After all, the *real* value of the work done by such departments is in the creation. The execution or manufacturing of the product is incidental. The same can probably be said of most of the rest of the departments to which audio visual functions report, including personnel, training and corporate administration. Only those audio visual departments that report directly to an engineering, data processing or manufacturing department might expect to find a sympathetic ear when it comes to matters of capital equipment. Thus, it is important for the audio visual manager to understand the mental outlook of the approving executive when preparing a capital request.

FACTORS AFFECTING THE CAPITAL BUDGET

Just as in operation and production budgeting, capital budgeting begins with a plan. The plan, however, should not begin with a "wish list" of what the audio visual staff would like to have. The plan instead begins with the work that will have to be done—what kinds of projects, in which media, with what frequency and at what cost. Other elements of the plan include the size and skills of available staff. It is of little value to propose the purchase of an animation camera if there is no one available to operate it.

The Company Capital Budget

The capital budgeting plan is also affected by the way the company treats such budgets. Manufacturing companies will normally have a company-wide capital equipment budget, which may be projected for two or more years into the future. All equipment, including audio visual equipment, becomes a part of that budget. One problem with this system is that the audio visual department finds itself competing for capital dollars with other departments, such as manufacturing and distribution, which may be able to make a more demonstrable contribution to the bottom line. On the plus side, however, the audio visual equipment budget is generally relatively small in comparison to the whole. An entire television studio can be built and equipped for less than the cost of one medium-sized stamping machine.

What constitutes a capital expense will also be defined differently in different organizations. Normally, capital equipment is any item that continues to have an asset value to the company for at least three years (the item must be retained for three years in order to qualify for the 10% investment tax credit in effect as of 1984). A minimum dollar value may also be set in order for the item to qualify as a capital asset, perhaps as low as $100, perhaps as high as $1000 or more. Anything that costs less may be written off in the first year as an operating expense. Of course, the depreciation on equipment that qualifies is an income tax deduction in most situations.

Some organizations may require that a capital budget be developed to cover all audio visual equipment purchased during the year. In other cases, only those items which exceed the predetermined dollar amount require budgeting. In yet other cases, if the total of all proposed purchases exceeds a predetermined dollar amount, then a budget is required. Some organizations may require approval by the board of directors if the dollar amount of a total annual purchase or of a single purchase order or item of equipment exceeds a specific sum. Clearly, it is essential for the audio visual manager to understand how the budgeting system works and, further, to understand how the purchase of capital equipment affects the company tax status.

Depreciation

Another factor affecting the approval of equipment purchases is the depreciation period established by the accounting department for each type of equipment. The actual life of the equipment may have little bearing on its tax life. If the company standard depreciation period is 11 years, it will not be getting the full depreciation value of a slide projector that will only last five years or an audio cassette recorder that will be lucky to make it for three years. I know of one company that treated its television studio (and all the equipment in it) as real estate and was depreciating it over 30 years.

When a piece of equipment does not last the full life, the balance of the depreciation write-down must then be taken in the year that the equipment is sold, traded in or junked. If the value of the equipment is significant, this can have a negative impact on the organization's long-term tax planning strategy. Therefore, wherever possible, it is suggested that a realistic depreciation schedule be assigned to each item or class of equipment at the time of purchase. How all of this affects the establishment of a capital budget will be seen shortly.

CAPITAL BUDGETING STRATEGIES

At one industry meeting an audio visual manager asked how he could go about establishing a capital budget. His problem was that he had no such budget; thus, each time he needed something he had to plead with his boss for permission to buy it. It is enough of a strain to have to do this once a year, much less 10 or 20 times a year, so I can well imagine what he was up against. It is worth noting that this manager worked for a service company, rather than one involved in manufacturing, and therefore a company which was probably not as familiar with the capital budgeting process.

The key to the capital budget lies in the plan. But it is incumbent on the audio visual manager to put on his or her management hat and develop the plan from the point of view of the benefits to the organization as a whole. Can it be proven that the purchase of a specific piece of equipment will contribute to the profitability of the company as a whole? How this is done can be seen from the capital equipment budgeting strategies which follow.

The Projected Cost Savings Strategy

The audio visual manager at the Pimlico Corporation had noted a significant increase in the amount of money being spent outside to produce slides for executive presentations. In fact, the cost had grown from $40,000 three years ago to $100,000 in the current year. This was partly because of increased costs for slide production but mostly because of increased demand. Further, the slides were becoming more complex and the deadlines tighter. A significant contributor to the increased cost was the amount of rush charges. The comparison he developed is shown in Table 9.1. If that trend were to continue at the present rate, next year a total of $140,125 would be spent to produce 2950 slides at an average of $47.50 each.

Table 9.1: Comparison of Annual Expenditures for Slides

	Number of Slides	Average Cost per Slide	Total
Three years ago	1,600	$25	$ 40,000
Current year	2,500	$40	$100,000

First it was necessary to establish whether this trend would, in fact, continue, or whether volume would hold at the present level or even decrease. To get a fix on that, the manager had to contact the clients for slide production and apprise them of the cost information to see what impact it would have. Sometimes just knowing the cost will cause a drop in volume. In this case, however, since the slides were all being done for top-level executives, cost was no object. Furthermore, there was no way that longer lead time could be allowed to reduce the rush charges.

The manager next explored alternate sources to see if lower prices could be obtained from vendors. He discovered he was already getting the lowest possible price that would still maintain the level of quality and service needed. Now he was free to develop the rationale for bringing that work in-house: a cost analysis that would be included in a capital equipment budget.

Cost Analysis

To develop the cost analysis, a number of different elements must be considered. The initial capital expense for the purchase of the equipment is calculated on the basis of the annual depreciation, which means that the life of the equipment must be calculated and, if the company depreciates all equipment on a standard term, that must also be considered. The calculation for the cost must include the following elements:

- The annual depreciation on the cost of the equipment;

- Any sales tax that must be paid at the time of purchase;

- Annual maintenance costs of the equipment;

- The income lost if that money were invested elsewhere, or additional money that may have to be borrowed by the company to purchase the equipment;

- The investment tax credit received by the company for purchasing this equipment; and

- Income tax benefits resulting from the write-off of depreciation on the equipment.

The AV manager at the Pimlico Corporation came up with the following information to use in his calculation:

1. The cost of the equipment was $155,000, including $78,000 for the graphics terminal, $53,000 for the film recorder and $24,000 for a film processor.

2. The sales tax was at the rate of 6%, or a total of $9,300.

3. He can expect a five-year life for the equipment, and the Pimlico accounting department indicated that a five-year write-down on depreciation would be reasonable. They also told him to use a straight-line calculation rather than differing percentages per year.

4. There was a 90-day warranty on the equipment, and thereafter an annual maintenance contract could be purchased for 8% of the value of the equipment. Since Pimlico had little experience with this equipment, it was decided to go with the contract for the purposes of calculation.

5. The accounting department told him to use 12% as the factor for the discounted cost of money over the five-year term.

6. At the time of the purchase, the investment tax credit was 10%, assuming the equipment was retained for at least three years. That credit was taken at the end of the first year.

7. The corporation was currently in the 50% tax bracket, meaning that 50% of each year's depreciation and 50% of the sales tax and annual contract maintenance charges could be deducted from the actual annual cost to the corporation. But the accounting department suggested that this not be included in the calculation, since the tax calculation was very complex and subject to too many variables.

Table 9.2 shows the calculation and the resulting costs of the equipment over the five-year term. The calculation provides for only three-quarters of the annual maintenance in the first year because of the 90-day warranty. For simplicity, the assumption is that the equipment is installed on the first day of the tax year and thus does not include any provision for prorating the depreciation, though that would be the more likely situation. A sales tax is also payable on the maintenance contract and is included under the tax column. The cost of money calculation is based on the entire $155,000 in the first year, since that net effect of the investment tax credit is not felt until the second year.

Non-Capital Costs

There are other costs associated with bringing this capability in-house. To handle the work, a computer graphics artist would have to be added to staff at a starting salary of $18,000 per year plus employee benefits at 35%. In addition, there would be the cost

Table 9.2: Calculation for Annual Cost of Equipment

	Depreciation (5-year)	Maintenance	Sales Tax (at 6%)	Cost of Money (at 12%)	Total Cost
Year 1	$27,900	$9,300[1]	$9,858[2]	$18,600[3]	$65,658
Year 2	27,900	12,400	744	14,880	55,924
Year 3	27,900	12,400	744	16,740	57,784
Year 4	27,900	12,400	744	16,740	57,784
Year 5	27,900	12,400	744	16,740	57,784
				Total Cost for 5 years	$294,934

[1]Assumes maintenance contract in effect for only nine months due to 90-day warranty.
[2]Sales tax on full cost of equipment payable in first year, in addition to sales tax on nine month maintenance contract.
[3]Cost of money calculated on full equipment cost in first year since net effect of investment tax credit not felt until year 2, which reduces cost of money factor in that year by calculation of investment tax credit of $15,500 x 0.12 = $1860.

of materials—film, chemicals, computer disks and slide mounts. This was worked out on the basis of total estimated cost divided by the number of slides anticipated and resulted in a figure of $1.25 per slide, including sales taxes.

Those are not fixed costs. Presumably the computer graphics artist would get pay increases during the period. These might be as high as 10%, with a similar increase in benefits. Inflation also affects the cost of materials, so that the $1.25 per slide cost would also increase. Past experience suggests that increases would be about 6% per year. And the number of slides produced could also be expected to continue to increase at 20% per year. Table 9.3 shows the total cost for equipment, personnel and materials, including the projected growth in volume and increases in salary and materials costs. A quick glance at even the first year internal costs of $93,826 shows a substantial savings over the projected outside costs of $140,125.

The next step is to compare the inside costs with the outside costs over the five-year period to determine just what kind of a return the company is getting on its investment. Table 9.4 shows that comparison, also assuming that the volume of work will continue to increase at the rate of 20% per year and that the cost of that work will increase at the rate of 6% per year.

The resulting savings over a five-year period will be $752,246. Even the total savings in the first two years—$137,193—are nearly enough to pay off the total initial investment. This is sufficient financial justification for making the investment in the in-house facility. There is also enough leeway here to make more conservative estimates on the volume increases. Reducing those projections to 10% per year would still show sufficient justification, though this would result in a somewhat longer payback period.

The Net Present Value Formula

There is, however, another way of calculating the resulting financial return to the

Table 9.3: Total Costs for Internal Production

	Equipment Costs	Personnel Costs (+20%/yr.)	Benefits	Total Personnel Costs	Units Produced (+20%/yr.)	Unit Cost (+6%/yr.)	Total Production Costs	Total Cost
Year 1	$65,658	$18,000	$6,300	$24,300	2,950	$1.25	$3,868	$93,826
Year 2	55,924	19,800	6,930	26,730	3,540	1.33	4,691	87,345
Year 3	57,784	21,780	7,623	29,403	4,248	1.40	5,966	93,153
Year 4	57,784	23,958	8,385	32,343	5,098	1.49	7,589	97,716
Year 5	57,784	26,354	9,653	35,578	6,117	1.58	9,653	103,015

Table 9.4: Projected Cost Savings for Internal Production

	Internal Costs				External Costs			
	Equipment	Personnel	Materials	Total	Units Produced (+20%/yr.)	Unit Cost (+6%/yr.)	Total	Net Difference
Year 1	$65,658	$24,300	$3,868	$93,826	2,950	$47.50	$140,125	($46,299)
Year 2	55,924	26,730	4,691	87,345	3,540	50.35	178,239	(90,894)
Year 3	57,784	29,403	5,966	93,153	4,248	53.37	226,720	(133,567)
Year 4	57,784	32,343	7,589	97,716	5,098	56.57	288,388	(190,672)
Year 5	57,784	35,578	9,653	103,015	6,117	59.97	366,829	(263,814)

corporation—the net present value formula. This formula projects the costs benefits of the future savings in today's dollars. This is a slight variation on the cost of money calculation that was used in Table 9.2, and yields a slightly different answer. In effect, it is the discounted value of the projected savings working backwards from the five-year point.

The net present value calculation disregards the depreciation on the equipment and instead balances the total cost of the initial investment ($155,000 less the 10% investment tax credit, plus the sales tax, or $148,800) and then compares that with the projected cost savings over the five-year period using an interest rate assumption. By using the same data regarding personnel and materials costs and the same assumptions with regard to increased volume and costs, as illustrated in Table 9.5, the resulting net present value is $798,099 at an interest rate assumption of 12%. This is a significant number, since in general even a net present value of $1 is sufficient to justify the investment from a financial standpoint.

The Pimlico AV manager was able to do all these calculations, including the net present value calculation, in less than an hour on his personal computer using a popular spreadsheet program, so the process was nowhere near as complicated as it looks. The formulas he used are reproduced in Table 9.6.

Even when the resulting payback is not as great as in this instance, being able to show resulting savings in such detail makes a very strong case. It shows that you've done your homework. However, even with such a clear-cut payback as this one, there would still be a problem with a company that required showing a one-year payback, and there are some that do. Requiring that any capital investment be paid back in one year severely limits such investments—which is undoubtedly the intention.

Different companies will ask that the depreciation or net present value be calculated over longer or shorter periods of time, ranging from 3 to 11 years. However, only the equipment depreciation will change; all other elements in the calculation will remain the same, including the cost of money.

But, as not all proposed purchases are going to show a direct payback on the investment, there are other strategies to consider.

The Maintenance of the Status Quo Strategy

About five years ago, the Pimlico Corporation had invested $100,000 in television production equipment. This was a one-shot request and approval. From time to time in the interim, the manager had made individual requests to replace equipment which broke down frequently and was expensive to repair. Sometimes the requests were approved, and sometimes not—then the manager had to continue to make do with what he had. Looking ahead, it was possible to see the time in the not too distant future when the entire system would grind to a halt as one piece of equipment after another broke down and might be irreparable due to obsolescence and a lack of spare parts.

Table 9.5: Net Present Value Calculation Using Spreadsheet Program

NET PRESENT VALUE CALCULATION

	Maint. (Inc. Sales Tax)	Materiel Units	Un. Cost	Sub.T	Personnel Salary	Benefits	Sub. T.	TOT. EXP.	External Exp. Units	Cost	Sub. T.	NET DIFF.
Year 1	$9,858	2950	$1.25	$3,688	$18,000	$6,300	$24,300	$37,846	2950	$48	$140,125	$102,280
Year 2	$13,144	3540	$1.33	$4,691	$19,800	$6,930	$26,730	$44,565	3540	$50	$178,239	$133,675
Year 3	$13,144	4248	$1.40	$5,966	$21,780	$7,623	$29,403	$48,513	4248	$53	$226,720	$178,207
Year 4	$13,144	5098	$1.49	$7,589	$23,958	$8,385	$32,343	$53,076	5098	$57	$288,388	$235,311
Year 5	$13,144	6117	$1.58	$9,653	$26,354	$9,224	$35,578	$58,375	6117	$60	$366,829	$306,454

Equipment Cost Calculation
Terminal	$78,000
Film Recorder	$53,000
Film Processor	$24,000
Total Cost	$155,000
Sales Tax	$9,300
Sub-Total	$164,300
Inv. Tax Credit	$15,500
Net Cost	$148,800

Gross Savings	$957,926
Net Present Value	$798,099

Table 9.6: Net Present Value Calculation Formulas

NET PRESENT VALUE CALCULATION

	Maint. (Inc. Sales Tax)	Materiel Units	Un. Cost	Sub.T	Personnel Salary	Benefits	Sub. T.	TOT. EXP.	External Exp. Units	Cost	Sub. T.	NET DIFF.
Year 1	@SUM(+C19*0.08)*0.75*1.06	2950	1.25	+C4*D4	18000	+F4*0.35	+F4+G4	+B4+E4+H4	+C4	47.5	+J4*K4	+L4-I4
Year 2	@SUM(C19*0.08)*1.06	+C4*1.2	+D4*1.06	+C5*D5	+F4*1.1	+F5*0.35	+F5+G5	+B5+E5+H5	+C5	+K4*1.06	+J5*K5	+L5-I5
Year 3	@SUM(C19*0.08)*1.06	+C5*1.2	+D5*1.06	+C6*D6	+F5*1.1	+F6*0.35	+F6+G6	+B6+E6+H6	+C6	+K5*1.06	+J6*K6	+L6-I6
Year 4	@SUM(C19*0.08)*1.06	+C6*1.2	+D6*1.06	+C7*D7	+F6*1.1	+F7*0.35	+F7+G7	+B7+E7+H7	+C7	+K6*1.06	+J7*K7	+L7-I7
Year 5	@SUM(C19*0.08)*1.06	+C7*1.2	+D7*1.06	+C8*D8	+F7*1.1	+F8*0.35	+F8+G8	+B8+E8+H8	+C8	+K7*1.06	+J8*K8	+L8-I8

Equipment Cost Calculation
Terminal	78000
Film Recorder	53000
Film Processor	24000
Total Cost	+C16+C17+C1
Sales Tax	+C19*0.06
Sub-Total	+C19+C20
Inv. Tax Credit	+C19*0.1
Net Cost	+C21-C22

Gross Savings	@SUM(M4..M10)
Net Present Value	+C23+@NPV(0.12,M4..M10)

Communication

The basis of the status quo strategy is a combination of a plan and communication. The communication part is relatively easy. It is nothing more than making the powers that be aware of the fact that no equipment can last forever. Over a period of time, every piece of equipment in the system will have to be replaced in order to keep that system operational.

This is a subtle process. It does not require any formal memorandums, just a hint dropped when the occasion presents itself. When visiting VIPs tour the studio, a casual comment in the presence of the vice president about how fortunate it is that such and such a piece of equipment has lasted so long and performed so well will serve to raise the level of consciousness.

The manager also started keeping track of all repair work on every piece of equipment. A file was kept, listing each piece of equipment, the reason for the breakdown, the amount of time it was out of service and the cost of repair. If the breakdown required that a substitute be rented or borrowed, that too was included in the report. The manager asked his technician to submit summary maintenance reports to him in writing every three months, with a copy to the vice president. The reports made special note of the costs involved and also noted any specific inconveniences that might have been caused by the equipment failure—such as the time the taping of the company president had to be postponed because the videotape recorder broke down.

Planning

As to the plan, the manager developed a table, listing every item of equipment and projecting the life expectancy of that equipment. Even though his company amortized everything on an 11 year basis, very few items in the television studio could be expected to last that long. Of course, he developed the table for more than just television equipment. It covered every item of equipment in the department. Based upon the date of purchase and the anticipated life expectancy, he was then able to project in what year that equipment would have to be replaced, and thus could lay out the projected capital budgets for the next five years for presentation to his vice president.

The question of projected life expectancy is, of course, open to a lot of discussion. There are no standard tables that tell us this, but through conversations with dealers, manufacturers and, most important, other users of similar equipment, a table can be developed.

Once the plan is presented, it is essential for the manager to monitor it continually. Some equipment may last longer than originally projected (as monitored by the maintenance records). Other equipment may have to be replaced sooner. Therefore, the five-year projection is clearly that—a projection—and should not become so ironclad that no deviation is possible.

Having opened a channel of communication and demonstrated that there was a specific plan to maintain the operation at its current levels made future discussions about replacing equipment much easier for our AV manager. Very few audio visual managers, however, would be content to simply maintain the status quo or upgrade capabilities only where there was a clear-cut cost benefit.

The Development and Growth Strategy

Just as any manufacturing operation must continually improve its ability to meet new demands and to develop new capabilities, so too must the audio visual operation strive to take advantage of new technology.

It is presumed that a need already exists in the work that is currently being done. It would be foolhardy to suggest buying equipment to meet a need which did not exist. Similarly, it is not advisable to attempt to convince the corporate executives that the installation of a new capability or the purchase of new equipment will generate a need. To propose buying a television digital effects system because, if it were available, people would use it is a technique that is almost certain to meet with not only rejection of the request but also a considerable loss of credibility for the audio visual manager.

The Valid Criteria

Nonetheless, there are three valid reasons for buying equipment under this strategy:

- Improved efficiency of operation;

- Convenience to the corporation;

- Reduction of outside expenses.

The three are interrelated—that is, the purchase of a specific item may meet all three criteria to one degree or another, though the purchase would not be justified if only one or possibly two of the criteria were applied.

For example the television production facility at Pimlico included a ¾-inch control track editing system. This was a two-machine editor capable of cuts-only edits. For some of the programs that were produced, A-B dissolves were necessary, which required the use of an outside editing house. The cost of this work—only a couple of thousand dollars a year—would not by itself justify the $30,000 investment required to upgrade the existing system to three-machine time-code A-B roll editing capability. So the manager looked to the criteria listed above.

Having this capability would, of course, save the outside charges. Since the outside editing house was frequently booked up well in advance, programs could also be completed more quickly with an in-house system. Further, it would save the time involved in traveling to and from the outside editing house for the program director and the client, who also attended the editing sessions.

In addition, when outside editing time had to be scheduled on short notice, the only available time was at night. This meant overtime for the staff producer and an especial inconvenience for the clients, both of which could be avoided with an in-house system. As an added benefit, the proposed time-code system was faster. Even a cuts-only program could be edited in 20% less time than was now the case with the control-track system.

There were other subjective benefits as well, such as more freedom to be creative in the production process, better quality and consistency of edits, and, at least in the mind of the manager, more effective programs. However, these are not criteria which are going to carry much weight in making a financial decision and are best left unspoken.

One way or another, all of the selected criteria do make a contribution to the bottom line, either directly or indirectly. It may be that the benefits for a proposed request do not add up to sufficient justification for the purchase. If that is the case, the audio visual manager should be the first to know.

Insufficient Justifications

Of course, a rationale based on improved efficiency can be carried too far. I overheard a conversation in which a request was being made for a new videotape editing system which was promised to reduce editing time by at least 30%. The executive being petitioned remembered being told that the purchase of the present system reduced editing time by 50%. "At that rate," he observed, "with the next generation of editing equipment, you'll be able to have the tapes edited before they're even shot." Rash promises about improved efficiency might also result in the suggestion that a staff reduction is in order, so the audio visual manager is cautioned to be realistic in making such projections.

The following are insufficient justifications for requesting new equipment:

- Imperceptible improvements in quality;

- Creativity;

- Another company has it;

- It's state of the art.

In developing the rationale for equipment budgets, the audio visual manager must always look at the request from the perspective of the company executive who must make the final approval. Again, the current economic climate and the condition of the company must also be a major factor. It simply doesn't make good sense to ask for capital budgets when the company is losing money.

Selling the Budget

Nevertheless, it is good practice to establish an annual audio visual capital budget, regardless of its size; an ongoing and continuing budget, based on a plan which, in turn, is based on a realistic assessment of the needs of the organization. It is far better to have an annual budget of only $5000 or $10,000 than to have to go back to explain to the approving executive all over again why you need to buy equipment in the first place.

Often, it is not so much the amount of money that is at stake, but the concept of the capital expenditure. In other words, if the audio visual manager is going to have to develop the rationale for the purchase of an animation camera, it might as well be a computer-operated one. In most equipment request situations, the actual dollar amount is not significant, since the capital expenses for audio visual equipment represent only a minuscule part of the total organization capital expenditures. If it is possible to get approval for any kind of animation camera, it is possible to get approval for the best one on the market with little or no additional effort.

In most situations, if the request is turned down, it is because the rationale is insufficient, not because the price is too high. Admittedly, we are in favor of buying the best equipment available to do the job. The saying "a carpenter is only as good as his tools" goes for a machinist, a computer operator or any job function that is equipment dependent. In general you get what you pay for. The more expensive the equipment, the longer it will last, the more efficient it will be, the lower the cost to maintain it, and the more productive the operator.

So, having spent several months developing his three-pronged capital equipment rationale, the Pimlico audio visual manager suggested to his vice president one day that it might be a good idea to establish an annual audio visual capital budget. Furthermore, he had developed a plan and some proposals and wanted to present the budget to the vice president in a couple of weeks. Without giving the vice president a chance to remind the manager that money was tight, that he couldn't see buying any more equipment right now, etc., etc., the manager went off to put the entire package together.

Two weeks later he walked in with the proposal. It called for spending $155,000 for the slide production system, $20,000 for phase one of the television studio equipment replacement, $30,000 for upgrading the videotape editing system, $5000 for replacements of other equipment (also based on the long-range plan) and a 10% contingency (since everyone knows that by the time an order for equipment is actually placed the cost will have gone up that much)—a grand total of $231,000. Attached to the package was the proposal to hire the computer graphics artist as well.

Alongside each item on the list was a brief summary of the supporting rationale, productivity increases, cost savings, time savings, convenience to the corporation and so on. However, there was not one word about more creative programming, keeping up with the state of the art or improved quality. Just hard facts. In fact, nowhere in the memorandum did it even say "we need this." Instead it stated only the clear benefits to the corporation.

Did the vice president deny or reduce the request? Being a vice president he probably did one or the other. But the proposal certainly made it much harder for him to do so. Most likely he looked at the $20,000 for the maintenance of the television equipment status quo and the $30,000 to upgrade the editing system and decided that since the editing system was part of the television facility the $20,000 maintenance budget could be reduced to $10,000, "for this year at least." Of course, the manager pointed out that any reductions in the maintenance equipment budget this year would result in an increased budget the next year.

LEASING

Leasing equipment does not really fall under the subject of capital budgeting, since leasing costs are treated as operating expenses for tax purposes. However, depending on the tax status of the organization, leasing can be a viable alternative to outright purchase, especially if cash flow is a problem.

As of early 1984, fixed-rate equipment lease rates were running between 12% and 13.5%. Floating-rate leases, where the interest rate varies, usually based on the prime lending rate, were running at 1% to 1.5% above the prime. The investment tax credit can either be taken by the borrower or kept by the lender. If the lender keeps the investment tax credit, up to 1.5 points can be shaved off the lease rate.

Leasing equipment will represent a higher total out-of-pocket cost to the lessor, since the rate paid will be above the prime interest rate, while we calculate the loss of interest on purchased equipment at the prime rate or slightly below it. After all, the lender is in business to make money too. But if the costs can be treated as operating expenses, there can be a significant tax savings.

The corporate tax status will have a great deal to do with whether or not leasing is economical. Many insurance companies, for example, are taxed not on the basis of net income but on the basis of insurance premium collected, so that neither lease nor purchase has any impact on taxes.

The projected life of the equipment will also be a factor. This will again depend on tax treatment, cash flow and a number of other esoteric financial considerations that are best left to the accounting department to sort out.

On the whole, however, any capital equipment purchase plan should also include consideration of the possibility of leasing, especially if a large amount of equipment is involved.

CONSTRUCTION CAPITAL BUDGETING

In addition to equipment, capital budgeting must also provide for the construction or renovation of facilities. Some of the dollar amounts quoted in the AVMA survey indicate that a number of companies are still building major facilities. I know firsthand of a few of recent vintage, such as those at J.C. Penney and Johnson & Johnson.

Budgeting capital expenditures for facilities is a lot more difficult than budgeting for the purchase of equipment. For one thing, the audio visual manager can't simply call up a contractor and ask how much it will cost to build a new audio studio. There are no published price lists for television studios or photo labs. In addition, facilities construction is usually long-term. It may take two or three years to complete a facility from the time an architect or designer is initially contacted. Just the process of developing a construction budget can take six months or more.

In some organizations, construction budgeting may be handled as part of a company-wide capital improvement budget. In the case of a new building, the cost of audio visual facilities will be included in the total cost of the building (almost impossible to break out as an individual item even if you wanted to). If there is any rule of thumb with construction costs, it is that the actual cost will be at least twice as much as you might reasonably estimate that it will be. Corporate construction usually requires adhering to a strict set of building codes, which are much more stringent than those for building a house. There are also very few houses that might need 1000 volts of electrical power or 30 tons of air conditioning, both of which are very high-ticket items.

Because of these complications, the audio visual manager will find it difficult, if not impossible, to do long-term capital construction budgeting. However, since any new facilities become part of the permanent plant, their cost is amortized over the life of the building, usually 20 or 30 years. Depending on the system in use at a specific company, it may be that the depreciation of these facilities is not directly charged to the audio visual department but is, instead, included in the overall rent rates for the building as a whole. It will be necessary for the audio visual manager to find out exactly how these costs are charged before suddenly finding that the annual budget has just been blown sky high when the amortization on the new photo lab is included.

In many cases, new equipment will incur construction costs. Installing an in-house slide production capability will require building a room for the computer terminal, with suitable electrical wiring and air conditioning. The cost of the plumbing for the film processor and even building an office or providing necessary physical space for the additional person must all be included in the calculation.

If these costs can be estimated, they can be included in the calculation, using the same formulas as are used in the projected cost savings strategy. The only difference is that the depreciation period will generally be longer.

CONCLUSION

In summary, capital budgeting is an emotional subject for audio visual people. Refusal to grant a request for a specific piece of equipment or a demand for reduction in the equipment request in general seems almost to be a personal affront to the audio visual staff. Unfortunately, many corporate executives tend to view audio visual people as equipment-happy gadget freaks. They find it incomprehensible that the cost of an animation camera can exceed the cost of a very expensive automobile or that a television camera can cost more than a medium-sized house. Since audio visual equipment requests are not usually directed to the same executives who buy mainframe computers or machine tools, there is little frame of reference. Until audio visual came along, the largest capital expense most such executives ever had to approve was for a new desk or electric typewriter.

Too often, audio visual managers and their staffs tend to reinforce that gadget-happy image. Equipment is requested that will improve the quality of a television picture, though only an engineer could see the difference. Certainly it could not be perceived by an eye as untrained as that of the executive who is being asked to approve the request. Equipment is requested that will produce more spectacular special effects, though no one so far had missed having them, except the AV people. Or equipment is demanded that can mount 300 slides a minute, meaning that an entire month's slide production can be mounted in 30 seconds. Such arguments are easily shot full of holes by organization executives, yet the audio visual people making these requests complain that "my boss doesn't understand me."

The problem is that the audio visual people are not being understanding or cognizant of their need to protect the company interest. A capital budget request based on a plan with sound rationale, well-presented and showing an understanding of the issues from the perspective of the individual who must approve the request, will be approved. The 81% of the companies in the AVMA survey with 1983 capital equipment budgets averaging $205,000 are proof of that.

Afterword

For those readers who have managed to stick with me thus far, I congratulate you. For those who skipped directly to this page hoping to find a quick summary of this book, I apologize. There is too much material, covering too broad a scope to consider attempting to summarize it in a few pages.

That there appears to be a demand for a book of this type indicates that audio visual is maturing. We have not gone as far as we can ultimately go in the development of new creative techniques or in the harnessing of new technology to meet our needs, but even as we continue to develop in those areas, we also realize the need to become more a part of our organizations. To do that means taking a long, hard look at the economics of our business. To that end, we are just beginning.

Early in this book I said that it was necessary to understand the business side of audio visual in order to make the transition from "technician" to "manager." I firmly believe that to be true. It is just as important to gain the trust and confidence of our companies for our managerial expertise as to earn their plaudits for our fine creative efforts.

Whether you believe that a chargeback system is good or bad, you may be forced to work within such a system, so it is well that you understand it. If you are among that fortunate group of managers who has input to the design of the budgeting system, then it is also important that you know as much as possible about the many systems available. In this way you can make the most informed, efficient choice for your corporation.

What we do will probably never be as essential to the business of our companies as, say, computers or telephones. Therefore, we must take every opportunity to demonstrate our commitment to the cost-effective use of the audio visual media. This can be done only through an understanding of the economics of audio visual production and the overall budgeting process.

If one thing has become evident from my research for this book, it is that the dollar investment in audio visual production and hardware is truly significant. If we assume that there are about 1200 major corporate audio visual departments, and if each has an operating budget averaging $822,000, then the total 1983 expenditure amounted to $986,400,000. Average capital budgets of $205,000 would add another $246,000,000. Corporate audio visual is indeed a $1 billion plus industry.

But those kind of statistics raise another concern. The fact is I did not find the level of financial expertise that I expected overall. Yes, there are many managers who are far more competent and knowledgeable than I in this area, but it also appears that an even larger number are relatively unsophisticated and demonstrate an alarming ignorance even of the budgeting methods within their own corporations.

If this book has been of some help in increasing your understanding of this difficult and arcane area, then it has been worth the effort. It is my intention to continue the development of this subject, and as I uncover additional information I will share it with my fellow professionals.

To that end, I have included a form inviting you to share your comments and to join the mailing list for future surveys. The form appears as the last page of this book.

Appendix A:
Source Directory

PROFESSIONAL ASSOCIATIONS

While a large number of professional associations may on occasion sponsor seminars specifically on audio visual budgeting and management subjects, the following organizations do so consistently in conjunction with their respective conferences.

Association for Multi-Image (AMI)
8019 N. Himes Avenue, Suite 401
Tampa, FL 33614
813-932-1692
Executive Director: Marilyn Kulp

Audio Visual Management Association
  (AVMA)
Box 821
Royal Oak, MI 48068
315-549-6585
Executive Director: Richard H. Joy
(Membership open to individuals with audio visual management and supervisory responsibilities in business and industry.)

International Television Association
  (ITVA)
6311 N. O'Connor Road, Suite 110
Irving, TX 75039
214-869-1112
Executive Director: Fred Wehrli

MAGAZINES

Most trade publications will carry articles on audio visual budgeting but the following do so with some consistency. While annual subscription rates are quoted, most are available at no charge to qualified readers.

Audio Visual Communications
United Business Publications Div.
Media Horizons, Inc.
475 Park Avenue South
New York, NY 10016
Monthly; $13.50/yr.

AVideo: The Total Communications Magazine
Montage Publishing, Inc.
2550 Hawthorne Boulevard
Suite 314
Torrance, CA 90505
Monthly; $18/yr.

E-ITV
C.S. Tepfer Publishing Company, Inc.
51 Sugar Hollow Road
Danbury, CT 06810
Monthly; $15/yr.

International Television: The Journal of the International Television Association
Business Publications Division
Ziff-Davis Publishing Co.
One Park Avenue
New York, NY 10016
Bi-monthly; $13/yr. (free to members of ITVA)

Millimeter, the Magazine of the Motion Picture & Television Industries
Millimeter Magazine
826 Broadway
New York, NY 10003
Monthly; $37/yr.

Photomethods
Business Publications Division
Ziff-Davis Publishing Company
One Park Avenue
New York, NY 10016
Monthly; $18/yr.

Video Manager
Knowledge Industry Publications, Inc.
701 Westchester Avenue
White Plains, NY 10604
Monthly; $24/yr.

PRODUCTION BUDGETING COMPUTER SOFTWARE

bt Systems Ltd.
137 E. 18th Street
New York, NY 10003
212-674-7480
Bidform AICP format production budgeting software

Comprehensive Video Supply Corp.
148 Veterans Drive
Northvale, NJ 07647
201-767-7990
Computer Aided Video software including *The Associate Producer, Power-Script, Datafax* and *Edit Lister*

Pacific Analog
4024 Dixie Canyon Avenue
Sherman Oaks, CA 91423
213-907-6029
AICP production budgeting software

The Production Source, Inc.
200 W. 58th Street
New York, NY 10019
212-765-4080
InstaBid custom and AICP format production budgeting and integrated accounting system

BOOKS

Very few books deal specifically with audio visual budget management; the first two books listed below do cover some aspects of budgeting. In addition, readers wishing to explore computer spreadsheet software programs will find the third book of interest.

Marlow, Eugene. *Managing the Corporate Media Center.* White Plains, NY: Knowledge Industry Publications, Inc., 1981.

Sambul, N.J., ed. *The Handbook of Private Television.* New York, NY: McGraw-Hill Book Co., 1982.

Henderson, T. and D.F. Cobb. *Spreadsheet Software From VisiCalc To 1-2-3.* Indianapolis, IN: Que, 1983.
(Comprehensive review of 10 electronic spreadsheet programs including *VisiCalc, SuperCalc, Calcstar, Perfect Calc, Supercalc 2, VisiCalc Advanced Version, Procalc, Multiplan, Context MBA* and *Lotus 1-2-3.)*

Appendix B: Budget Management Forms

The following collection of budget management, control and reporting forms is not intended to be comprehensive. It simply illustrates the wide variety of forms that can accomplish more or less the same task. Additional forms appear in Chapter 5.

The appendix is divided into three parts. Part 1 includes forms used in the development and reporting of annual departmental operating budgets (Figures B.1-B.5). As is the case with all of the forms in this appendix, some are provided by the company as a standard form to be used by every activity. Others are specifically designed by the AV manager to meet the primary objective of providing a standardized and consistent method of budgeting.

Part 2 is devoted to forms used in developing budgets and obtaining client approval for specific projects (Figures B.6-B.13). Included in this group are project budget worksheets as well as formal estimates or requisitions to be sent to the client to get a "sign-off" of the budget commitment.

Part 3 consists of forms used to report actual expenses: time, materials and outside services (Figures B.14-B.25). These forms are usually used as input briefs for a computer system that will actually calculate and bill the charges as indicated by code numbers. However, they may just as easily be used to develop a manual project report or simply be kept on file to provide a permanent record of the actual time or expenses involved in the specific project.

Many managers tend to feel that if they have enough forms, they will have solved all of their problems. But as experience will show, it's not the form but what you do with it that counts. The information derived from the form is only as good as the use that is made of it. In recent years there has been a tendency to reduce the amount of paperwork (as evidenced even by the federal government's Paperwork Reduction Act). This has been helped considerably by the advent of computers and spreadsheet programs, as discussed in Chapter 8.

A final caveat: in designing or developing forms for internal use, the AV manager is well advised to remember that every form means that someone has to take the time to fill it out.

PART 1: FORMS FOR DEPARTMENTAL OPERATING BUDGETS

Figure B.1: A Budget Forecast Form for Equipment Expense

AV Communications
1984 Equipment Expense (Maintenance) Budget Forecast   Date:_____

Page:_____

	1st Quarter	2nd Quarter	3rd Quarter	4th Quarter	Annual
1983 actual & projected.					
1984 projected (list major items by quarter)					
Totals					
% Change – 1984 over 1983					

Figure B.2: A Company-Wide Standard Proposed Budget Form

ACCOUNT	DESCRIPTION	BUDGET	ESTIMATED ACTUAL	PROPOSED		BUDGET	ACTUAL	PROPOSED

COMPANY ___   PROPOSED BUDGET   YEAR ___   DEPARTMENT ___   COST CENTER ___

NUMBER OF PERSONNEL

DISTRIBUTION

Figure B.3: A Table of Weeks for Biweekly Pay Periods

1984
TABLE OF WEEKS - BIWEEKLY PAY PERIODS   FORM TW

1983				1984					
Date of Hire or Increase	No. of Wks in Qtr.		To End Of	Date Of Hire or Increase	Number of Weeks in Quarter				To End Of 1984
	Third	Fourth			First	Second	Third	Fourth	
				January 1	12	14	12	14	52
				January 15	10	14	12	14	50
				January 29	8	14	12	14	48
				February 12	6	14	12	14	46
				February 26	4	14	12	14	44
				March 11	2	14	12	14	42
				March 25	0	14	12	14	40
				April 8	0	12	12	14	38
				April 22	0	10	12	14	36
				May 6	0	8	12	14	34
				May 20	0	6	12	14	32
				June 3	0	4	12	14	30
				June 17	0	2	12	14	28
				July 1	0	0	12	14	26
				July 15	0	0	10	14	24
July 31	8	14	22	July 29	0	0	8	14	22
August 14	6	14	20	August 12	0	0	6	14	20
August 28	4	14	18	August 26	0	0	4	14	18
September 11	2	14	16	September 9	0	0	2	14	16
September 25	0	14	14	September 23	0	0	0	14	14
October 9	0	12	12	October 7	0	0	0	12	12
October 23	0	10	10	October 21	0	0	0	10	10
November 6	0	8	8	November 4	0	0	0	8	8
November 20	0	6	6	November 18	0	0	0	6	6
December 4	0	4	4	December 2	0	0	0	4	4
December 18	0	2	2	December 16	0	0	0	2	2

Figure B.4: A Salary Worksheet

Form SW

1984
SALARY WORKSHEET

Function _____

☐ Exempt   ☐ Non-Ex.   ☐ Part Time

Pay Weeks Per Quarter

Name	Date	Start	Weekly-In-Force Inc.	1983 Add.	1984	1st Quarter 12	2nd Quarter 14	3rd Quarter 4/12	4th Quarter 14/14	TOTAL

TOTALS

Figure B.5: A Travel, Entertainment and Moving Worksheet

FORM TEM

TRAVEL, ENTERTAINMENT AND MOVING WORKSHEET

TRAVEL AND ENTERTAINMENT

	Transportation	Meals and Lodging	Entertainment	Personal Professional Memberships	Other Expense	Total
Regular Travel and Entertainment Accounts: 0100						
TOTAL						

MOVING

MOVING	Real Estate Reimbursement	Relocation Bonus	Other Moving	Total
Moving Expense Account: 0120				
TOTAL				

02-9981-35D

PART 2: FORMS FOR SPECIFIC PROJECT BUDGETS

Figure B.6: An Audio Visual Program Request Form

| AUDIOVISUAL PROGRAM REQUEST | | | | FORWARD TO: | TV-2 |

| COMPANY | LOC. CODE | EXPENSE CODE | DEPARTMENT | REQUESTED BY | DATE |

| DESCRIPTION | | EXT. NO. | STA. NO. | DATE NEEDED |

Suggested Medium: ☐ Video ☐ Audio ☐ Slides ☐ Slide/Sound ☐ Transparencies

Creative Information

Number of copies needed

AUDIOVISUAL SECTION USE ONLY

| MATERIALS | SERVICES |

| PRICE QUOTED | DATE | INITIALS |

LABOR	HOURS	
	STAFF	CLERICAL
WRITING		
ARTWORK		
PHOTOGRAPHY		
PRODUCTION		
RECORDING		
PRESENTATION		
EDITING		
DUPLICATION		
AUTHOR'S ALTER		
TOTAL HOURS		
COST		

PRODUCTION COMPLETED (date & init) _____

APPROVAL COPY SENT   (date & init) _____

APPROVED DATE   (date & init) _____

DUPLICATION COMPLETED (date & init) _____

PROJECT COMPLETED   (date & init) _____

LABOR   _____

MATERIALS   _____

SERVICES   _____

TOTAL COST   _____

Job Number_____ Assigned To:_____

Figure B.7: A Request Form for Graphics Materials

MEDICAL GRAPHICS

Requested By | Artist Name: Computer: ☐ | Date Received | Section | Date Due | Date Completed | Project Number

*Staff Approval

TITLE: | Account Number | Grant or SPF No. | Accounting Use Only

GENIGRAPHICS   Diskette ID No. & Other Information

DESCRIPTION	TYPE	ID	QUANTITY		DESCRIPTION	TYPE	ID	QUANTITY
Base Charge 1	110		.00		Word Slide 1	330		.00
Base Charge 3	130		.00		Word Slide 2	335		.00
Graph 1	310		.00		Table 1	340		.00
Graph 2	320		.00		.Table 2	345		.00
Diagram 1	350		.00		Title Slide	325		.00
Diagram 2	360		.00		Consultant	200		hrs
Minor Revision	290		.00		Operator Time	300		hrs
Duplicate Slide	450		.00		Special Charges	900	50	.00
Boardwork	301		hrs		Conversions	920		.00
Pos-one exposure	410		.00		Conceptualizing	990		hrs

OTHER   (Project Description)

DESCRIPTION	TYPE	ID	QUANTITY		DESCRIPTION	TYPE	ID	QUANTITY
Base charge - 1	110		.00		Color pos exposure	412		.00
Base charge - 2	125		.00		Pos-one screening	414		.00
Base charge - 3	130		.00		Color key transparency	416		.00
Base charge - 4	150		.00		Color key opaque	418		.00
Base charge - 5	160		.00		Color key screening	420		.00
Line - 1	210		.00		3M image and transfer	422		hrs
Line - 2	220		.00		Typesetting	424		hrs
Line - 3	230		.00		Paste-up	426		hrs
Tone - 1	240		.00					hrs
Tone - 2	250		.00		Backpainting/coloring	430		hrs
Sketches	260		.00					hrs
Comprehensive	270		.00		Medical illustrator	200		hrs
Minor revision	290		.00		Scientific illustrator	300		hrs
Graph - 1	310		.00		Graphics designer	400		hrs
Graph - 2	320		.00					hrs
Diagram - 1	350		.00		Medical sculptor	500		hrs
Diagram - 2	360		.00		Exhibit fabricator	600		hrs
Pos-one exposure	410		.00		Materials	800	50	
					Special charge	900	50	

©: ☐ Yes   Base Color No. _____ Copied in Medgr. ☐

ARTWORK ID NUMBER(S)   Use Back of Last Sheet for Pt No.

KEYWORDS (Use Back of Last Sheet)

SPECIAL INSTRUCTIONS

PHOTOGRAPHY   No. Artwork Pieces

	AUTHOR No. Each	B & W No. Each	MEDGR	AUTHOR No. Each	COLOR No. Each	MEDGR	AUTHOR No. Each	DIAZO No. Each	MEDGR
SLIDES Date Needed ____	Total No.	No. Each		No. Each	No. Each		No. Each	No. Each	
PRINTS Date Needed ____	Total No.	No. Each		No. Each	No. Each		No. Each	No. Each	

REVISIONS: ☐ SAVE ☐ DESTROY OLD NEG

MC 1699/R481

Figure B.8: A Planning Sheet for Production Needs and Costs

Production # :
Department:

Requested by:
Taping Date:

Date:
Est. Time:
Comp. Date:

Functions	Availability	Completion Date	Responsible	Cost Per	Time/Quan.	Total
PRODUCTION						
In Studio						
On Location						
Outside Fac.						
					Sub Total:	
EQUIPMENT/POWER REQ.						
Studio Equip.						
Portable Equip.						
Rental Equip.						
Power Needs						
					Sub Total:	
PREPRODUCTION						
R & D						
Script/Outline						
Sec. Services						
Printing						
Graphics						
Photos/Film/Slides						
Tape						
Sets						
Props						
Costumes						
Lights						
Sound Track						
					Sub Total:	
POSTPRODUCTION						
Tape Review						
Editing						
Tape						
Video Dubs						
Audio Dubs						
TBC						
Duplication						
Packaging						
Shipping						
					Sub Total:	
PERSONNEL						
Talent (Prof.)						
Prod./Director						
Crew						
Other						
					Sub Total:	
TRAVEL						
Transportation						
Lodging						
Meals						
Miscellaneous						
					Sub Total:	
					Grand Total:	

Figure B.9: A Project Analysis Worksheet

PROJECT ANALYSIS WORKSHEET   (Trial Form - 1 Feb 80)   Page 1
Contact Name_____Phone_____
Department _____Location_____
Working Title_____Proj #_____
Date of Original Contact_____Project Mngr/PD_____
FINAL PROGRAM TITLE_____
FINAL CATALOG #s/TIME(s)_____

CLIENT INTERVIEW Date_____Attendees_____
1. Describe your primary audience and potential secondary audiences._____
_____

2. Describe existing situation, need, attitude, status needing change thru communications.
_____
_____

3. Without assuming any particular medium, what kind of change, result, response, status or
improvment would you consider a successful resolution to Question 2's answer?_____
_____
_____

4. Is there a deadline?_____
5. Are there budget limitations?_____$_____
6. Describe how audience will receive/view/have access to program; e.g. class, library, mail,
at conference, seminar, etc._____
_____

7. From whom and/or what will resource material for program come?_____
_____
_____

8. Are there special talent and/or graphic requirements?_____
_____
_____

9. List Subject Matter Key Words for Catalog Index._____
_____

10. Other _____
_____
_____
_____

A. Content/Budget Approval Authority (AVP)_____
B. Program Funding Source (LC, NLC, OTHER) C. BDP Code for Tape Surcharge _____
D. Estimated Value/Savings _____$_____
E. Feedback/Effectiveness Measurement_____
F. Viewing Restrictions_____
G. Estimated Program Lifetime_____
H. Appropriate for Secondary PR Use?_____
I. Content Category (Policy, System Planning, Strong Recommendation, Recommendation, Info)
J. Applicable PR Objectives_____

REVIEW DUPLICATION/DISTRIBUTION/PROMOTION INTERVIEW QUESTIONS ON PAGE THREE----------------

(Use this area to make suggestions for improving this trial form)

(continued on next page)

Figure B.9: A Project Analysis Worksheet (cont.)

SUPPORT SERVICES/PERSONNEL/LOGISTICS BUDGET RECORD   Page 2
    Item   Supplier(s)   Proposed   Actual
1. Writer(s) $ $
2. Graphics
3. Scenery
4. Props
5. Tech Rentals
6. Music
7. Photography Stills
8. Photography Motion
9. Useage Fees
10. Shipping
11. Travel
12. Catering/Meals
13. Lodging
14. Research/Consultation
15. Casting
16. Narrator
17. Talent

18. Freelance

19. Other

20. Editing
21. Audio Mix
22. Contingency_____ % Re-Edit, Over-Time, Equipment Failure, etc.

23. TAPE SURCHARGE (For Production, Client & Library Tape Copies) $200 $200

    PRODUCTION SUB-TOTAL -- $ $

24. Tape/Film Duplication & Distribution
25. Film Transfer
26. Promotion Writer/Photography
27. Other Post-Production

    POST-PRODUCTION SUB-TOTAL --------------------------------------- $

                                  TOTAL BUDGET $

PLANNING CHECKLIST
a. Get AVP Budget/Treatment Approval_____   i. Is a Remote Survey Needed
b. Take Promotion Photos/Write Copy_____   j. Is a Talent Briefing planned
c. Are Non-Disclosure Forms Required_____   k. Has Engineering been briefed
d. Is Legal Review Required of Content_____   l. Have you reviewed facilities bookings
e. Is Feedback/Evaluation planned_____   m. Is Network Booked ____ · TWX written
f. Is a PR Module applicable_____   n. Have graphics spelling been checked
g. Is Legal Review Required of Contract_____   o. Is Prompter script in format
h. Review Compliance with Talent Union_____   p. Get text for Cassette Labels

(continued on next page)

Figure B.9: A Project Analysis Worksheet (cont.)

FACILITIES PLANNING & RECORD DATA   Page 3
Date   Facilities/Equipment/Rentals/Circuits/Special Needs Performance/Trouble History

_____
_____
_____
_____
_____
_____
_____
_____

TAPE USE 1"_____Reels   TIME LOG Prod/Editing _____ Hrs
        3/4"_____Cassettes   Overtime   _____ Hrs
        Audio _____Reel/Cass   Down Time   _____ Hrs

LIVE NETWORK TELECAST DATA
SSO#_____ XMISSION DATE/TIME_____ Q & A ?_____
TOPIC #_____ LOCAL BUILDING FEEDS?_____
Q&A? Yes/No Q&A METHOD _____ NUMBER ON CAMERA_____
CHECKLIST:
A. Review Network Points & Conf Call List ___ E. Should Audio Tape/Transcript be Made?___
B. Prepare TWX Copy for Field Notification ___ F. Are Cassette Copies needed Immediately__
C. Can Program be Recorded in the Field?   ___ G. What If Network is Cancelled?   ____
D. Will telecast include videotape?   ___ H. Brief talent on Network Procedure   ____

DUPLICATION/DISTRIBUTION/PROMOTION/FEEDBACK INTERVIEW QUESTIONS
1. Review duplication/distribution procedures and costs...Tentative copy order is_____
2. How many copies needed immediately following final edit?_____
3. Will a letter or other material accompany tapes when distributed?_____
4. Is there an address mailing list for videotape distribution?_____
5. Will client or addressees pay for videocassette copies?_____
6. Has videocassette label copy been written and approved?_____
7. How will future copies of program be ordered?_____
8. Review promotion outlets for client approval (Bldg Channels, Newspaper, Poster, Handouts)
9. Review promotion & distribution plans for PR module._____
10. Is Film Transfer a possibility for program?_____ Is Super 8 a possibility?_____
11. Does the client have a feedback or program effectiveness measurement for program?_____
12. Discuss and choose a methodology for determining program effectiveness._____

13. Review element reels recycle policy and PR-IRS catalog system,_____
14. Review document retention procedure for script or transcript,_____
15.

POST-PRODUCTION ACTION CHECKLIST   Date   Date
A. PR-IRS Form Submitted ____/____/____ E. Feedback Report Completed___/___/___
B. Element Reels Recycled ____/____/____ F. Program Revised on ____/____/____
C. Dub Master to Duplicator___/___/___   G. Program Declared Obsolete on ___/___/___
D. Follow-Up Letter to Client ___/___/___ H. Project Cancelled on___/___/___

ADDITIONAL NOTES:_____
_____
_____
_____
_____
_____

DESCRIPTIVE NARRATIVE OF COMPLETED PROJECT   Page 4
Write summary of final program including key subjects covered, key personalities, general
technique used in production, etc. This material can be used for or taken from promotional
program description. Scripts, still photos, bills, memos, receipts or any other records of
the program should be filed with this finished PROJECT WORKSHEET.

Figure B.10: An Estimating Worksheet for Conference, Multi-Image and Slide Projects

```
AUDIO VISUAL SERVICES DIVISION                          DATE_____

CONFERENCE/MULTI-IMAGE/S/SLIDE ESTIMATE WORKSHEET       PROJ. #_____

ASSIGNMENT_____: PRELIMINARY_____: REVISION_____   MEDIA_____

TITLE?EVENT_____   DIV. CODE_____

DIVISION_____ CONTACT_____  TEL. #_____

PRODUCER_____ TEL. #_____

# SLIDES_____ # COPIES_____

*****************************************************************************************
INTERNAL CHARGES
                        No.        Unit        Item
Staff Time              Units      Cost        Cost

Producer (Hrs/Days)     _____    $ 60/425     $_____
Technician (Hrs)        _____      35         _____
Projectionist (Hrs)     _____      25         _____
Estra Crew (Hrs)        _____      25         _____         TOTAL STAFF
Artist (Hrs)            _____      50         _____
Photographer (Hrs/Days) _____      50/300     _____         $_____
Type I Slides (Rush)    _____      25 (40)    _____
Type II Slides (Rush)   _____      40 (60)    _____
Type III Slides (Rush)  _____      60 (100)   _____
Builds                  _____      10 (15)    _____

Facilities

Audio Studio (Day/Night) _____     800/90     _____
Audio Dubbing/Dupe (Hr)  _____     75         _____
Audio Edit/Search        _____     25         _____
Multi-Image System(Day)  _____     1,700      _____
Conference Audio (Day)   _____     1,000      _____         TOTAL FACILITIES:
Equipment Rental         _____     _____      _____
Other_____        _____     _____      _____         $_____
Contingency.................................................     $_____
                                          TOTAL INTERNAL CHARGES $_____
EXTERNAL CHARGES

Materials

Film (Rolls/Frames)      _____     $10/.30    $_____
Additional Proofs        _____     2/set      _____
Slide Dupes              _____     2          _____
Slide Trays              _____     10         _____
Slide Mounts (Glass)     _____     1          _____
Audio Tape               _____     _____      _____
Audio Tape               _____     _____      _____
Audio Cassettes          _____     _____      _____         TOTAL MATERIALS:
Audio Cassettes          _____     _____      _____
Other_____        _____     _____      _____         $_____

Outside Purchases and Expenses

Professional Talent__Actors for___ days       $_____
Professional Services                         _____
Licenses - Music/Stock Photos                 _____
Genigraphics Charges                          _____
Equipment Rental                              _____
Staging/Security/on-site crew                 _____
Shipping                                      _____

Travel

Fares $_____X_____ people X____trips        $_____
Per Diem $125/day for___trips                  _____         TOTAL PURCHASES:
Other_____          _____

Administrative Service (10% or max. $200/item)  $_____        $_____

                                          TOTAL EXTERNAL CHARGES   $_____
Comp. Date_____
                                          TOTAL ESTIMATED COST     $_____
```

Figure B.11: A Funding Authorization Form

FILM/VIDEO PRODUCTION FUNDING AUTHORIZATION

JOB NUMBER	DATE OF ESTIMATE	PROJECT START DATE	DATE PROJECT DUE

REQUESTOR NAME	DEPT. NAME	CHARGE NUMBER	DIV. REF. NUMBER

JOB TITLE	CLASS	SALES CAT.

PRODUCER	FORMAT	PURPOSE

JOB SPECIFICATIONS	QUANTITY

ESTIMATED COST (INCLUDE VENDING)

DEVIATIONS TO ORIGINAL REQUEST

COST OF DEVIATIONS

APPROVALS

COMMUNICATIONS REPRESENTATIVE	DATE	REQUESTER	DATE
COMMUNICATIONS MANAGER*	DATE	REQUESTERS MANAGER*	DATE

*SEE AUTHORIZATION PROCEDURE FOR SIGNATURE LEVEL

HF-341   **DISTRIBUTION:**   WHITE - CUSTOMER - SIGN AND RETURN TO:

BLUE - CUSTOMER KEEP
GREEN - JOB FILE
YELLOW - PROFIT SYSTEM
PINK - LIBRARY
GOLDENROD - PRODUCER

Figure B.12: A Request for Photographic Services

REQUEST FOR PHOTOGRAPHIC SERVICES

AV AUDIO VISUAL SERVICES

Requested by . . . .

Date of Requisition

Floor and Building . . . .

Charge to (Division) . . . .

Date Finished Work Req'd

Division Code ☐☐☐ or AV Project No. ☐☐ ☐☐☐

TYPE OF WORK
☐ Photographer
  Date Photog. Req'd ☐☐ ☐☐ ☐☐
☐ Reprints
☐ Copy
☐ Other(describe below)

FINAL PRODUCT
☐ Contact Sheet
☐ Color Transparencies
☐ Color Prints
☐ Black & White Prints
☐ Polaroid

Common Print Sizes
4x5
5x7
8x10
11x14
16x20
Other sizes available upon request

BRIEF DESCRIPTION OF SITUATION TO BE PHOTOGRAPHED. If this is a request for reprints from negatives in our files, indicate the negative number, and the size(s) and amount of prints required. If this request is for portraits from our files, indicate the name and negative number. To expedite your request, please complete the form where required.

Photo Section Use Only
Phot. _____
BW35 _____
BW35-P _____
BW120 _____
BW120-P _____
BW4/5 _____
BW8/10 _____
VPS35 _____
VPS120 _____
VPS4/5 _____
K64 _____
EK35 _____
EK35-P _____
EK120 _____
EK120-P _____
EK4/5 _____
EK8/10 _____
POL. _____
_____

PLEASE NOTE: The level of approval will depend upon the dollar amount involved and the policy of your department. Unless you are sure, it is best to check with your departmental budget coordinator to make certain there is money available in the EMIS 277 (Audio Visual Services) budget for your request.

Name/Approved By:

. . . .

DO NOT USE OTHER SIDE

Comb 82408 Ed 3-80   SEND TO: Photo/Graphics Section, Audio Visual Services, 13 Gib.   Cat #731867M

(continued on next page)

Figure B.12: A Request for Photographic Services (cont.)

Form fields (top row): PERS | DATE OF WORK | WORK CODE | C | TIME | PROJECT CODE | DIV CODE | CLIENT CONTACT NAME

Form fields (bottom row): ITEM CODE | DATE CHARGED | PERS | C | UNIT | PRE CHG | PROJ CODE | DIV CODE | CLIENT CONTACT NAME

PHOTOGRAPHIC CODES:

B&W	Color	
400	500	Film 120/35
401	501	Film 4x5
402	502	Film 5x7
403	503	Film 8x10
404	504	Film 11x14
405	505	Film (by frame)
406	506	Film Only (Neg)
	507	Film Only (Sld)
	508	Process Only (Sld)
410	510	First Print 5x7
411	511	Add'l print 5x7
412	512	First Print 8x10
413	513	Add'l print 8x10
414	514	First print 11x14
415	515	Add'l print 11x14

B&W	Color	
416	516	First Print 16x20
417	517	Add'l print 16x20
418	518	Polariod
	519	Passport
	520	Internegative
421	521	Copy negative
	530	Duplicate slides . . . 1st Dupe
	531	Duplicate slides . . Each add'l
440	540	Mount/illustration board
441	541	Mount/foam core
450	550	Add'l contact
	560	Dupe trans 4x5
	561	Dupe trans 5x7
	562	Dupe trans 8x10
	563	Dupe trans 11x14

ADDITIONAL CODES:
600 Glass Mount
602 Soft Slide Sheet
603 Hard Slide Sheet

TIME CODES:
130 Photographer Hour
132 Photographer Day
192 Machine Maintenance

PREMIUM CHARGES:
4.00:B&W Photo Same Day Service
2.00:B&W 24 Hour Service
2.00:Color 72 Hour Service

Figure B.13: A Cost Worksheet Showing Estimated and Actual Hours and Costs

AUDIO VISUAL TASK COST WORK SHEET X1039 (REV. 10-74)	PROJECT NAME		PROJECT NUMBER	
PROJECT STARTED		PROJECT COMPLETED		
COST ITEM	ESTIMATED HOURS	ACTUAL HOURS	ESTIMATED COST	ACTUAL COST
AUDIO VISUAL SERVICES PERSONNEL (SPECIFY)				
NARRATORS/ACTORS				
MATERIALS CONSUMED				
FILM/TAPE STOCK				
GRAPHIC ART SERVICES				
OUTSIDE AUDIO SERVICES				
OUTSIDE STUDIO RENTAL				
OUTSIDE STUDIO PERSONNEL				
PHOTOGRAPHIC SERVICES - STILLS				
- MOTION PICTURE				
- AERIAL				
- ANIMATION				
OUTSIDE ART SERVICES				
OUTSIDE SCRIPT SERVICES				
EQUIPMENT RENTAL				
FILM PROCESSING AND PRINTING				
POST PRODUCTION				
OUTSIDE EDITING				
A & B ORIGINAL FILM CUTTING				
OPTICAL PRINTING/VISUAL EFFECTS				
MUSIC SELECTION & PRE-RECORDING				
AUDIO MIX & INTERLOCK PROJECTION				
OUTSIDE TELECINE				
TRAVEL AND EXPENSES				
TRANSPORTATION/FREIGHT CHARGES				
OTHERS (LIST AND SPECIFY)				
TOTALS				

COMMENTS ON PROJECT

PROJECT ☐ ON TIME   ☐ LATE

PART 3: FORMS FOR REPORTING ACTUAL EXPENSES

Figure B.14: An Hourly Rate Billing Sheet

Date Received: _____

Production Date: _____

Date Promised: _____

Job Description: _____

Client's Name: _____

Address/Department Code: _____

Date	Name	Service	Hours	Rate	Cost

Total Man-Hours: _____

Total Man-Hour Cost: _____

Figure B.15: A Manpower Record Form

VIDEO TAPE PRODUCTION   MAN POWER RECORD

VIDEO TAPE TITLE _____

TV ACTIVITY															MAN/HR TOTAL
MEETING															
SCRIPTING															
SCRIPT BLOCKING															
VISUALS															
REMOTE															
STUDIO															
CONTROL ROOM															
DUBBING															
PREVIEW & EVALUATION															
CUSTOMER SCRIPTING, VISUALS, ETC															
OTHER															

Figure B.16: A Miscellaneous Work Order Form

FILM/VIDEO SERVICES
MISCELLANEOUS WORK ORDER

DATE	REQUESTOR		DEPARTMENT NUMBER	
RECEIVED BY			DATE REQUIRED	
ASSIGNED TO				
DESCRIPTION OF PROJECT				
MATERIAL USED		MATERIAL CHARGED		
DATE	BY			
DELIVER TO	VIA		DATE DELIVERED	
	REMARKS			

HM–407

Figure B.17: A Job Process Record Form for Audio Visual Materials

Week Ending: ____ MO. DAY YR.

AUDIO VISUAL JOB PROCESS RECORD
AUDIO VISUAL MATERIALS

EXPENSE CD (1) ___ (7) LOCATION CD (8) ___ (12) JOB NUMBER (13) ___ (16) TYPE CODE (17)(18)(19) 0 1 2 DESCRIPTION ___ (33)

AUDIO VISUAL MATERIAL	QUANTITY	UNIT COST	TOTAL COST
Hi-Con-Slides			
35 mm Color Film			
35 mm Color Film Processing, per roll			
35 mm Color Film Processing RUSH Processing, per roll			
Audiocassettes			
Audio Tape (reel to reel)			
Transparencies			
¾″ Videocassettes			
½″ VHS Videocassettes – 1 HOUR			
½″ VHS Videocassettes – 2 HOUR			
Art Materials			
Miscellaneous:			
TOTAL AV MATERIALS		(34)	(46)

FORM 4960 NS (1/83)

Figure B.18: A Job Process Record Form for Outside Audio Visual Services

AUDIO VISUAL JOB PROCESS RECORD
AUDIO VISUAL OUTSIDE SERVICES

Week Ending: _____ MO. DAY YR.

EXPENSE CD (1)____(7)   LOCATION CD (8)____(12)   JOB NUMBER (13)____(16)   TYPE CODE (17)(18) 0 3   DESCRIPTION

AUDIO VISUAL OUTSIDE SERVICES	QUANTITY	UNIT COST	TOTAL COST
Announcers			
Studio Rentals			
Actors			
Equipment Rentals			
Labels			
Slide Dupes			
Camera Operators and Technicians			
Music Licenses			
Props and Furniture			
Outside Suppliers			
Miscellaneous:			
		(34)	(40)
TOTAL OUTSIDE SERVICES			

FORM 4959 NS

Figure B.19: An Input Sheet for Special Material and Services Purchased Outside

INPUT SHEET
SPECIAL MATERIAL AND
OUTSIDE PURCHASED SERVICES

EMPLOYEE NUMBER
900

START TIME
8:00

SPECIAL MATERIAL AND OUTSIDE
PURCHASED SERVICES

DATE

PAGE

COST CENTER AND OPERATION
CODES ARE LISTED ON THE REVESE OF
THIS SHEET

JOB NUMBER	SPECIAL CODE A B C	COST CTR	OPER NUMBER	QUANTITY	MATERIAL/PURCHASE COST	INVOICE NUMBER	REQUESTER NAME

(continued on next page)

HS 596 REV 7-83

Figure B.19: An Input Sheet for Special Material and Services Purchased Outside (cont.)

SPECIAL MATERIAL (M)

57 PRESENTATION PREMIUMS/MODELS
570 PREMIUMS/MODELS

58 PRESENTATION MANAGEMENT SUPPORT
580 SPECIAL GIFTS

81 PAPER
810 UNCOATED PAPER
811 COATED PAPER
812 UNCOATED COVER
813 COATED COVER
814 BOND
815 LEDGER
816 INDEX
817 CARBONLESS
818 ENVELOPES
819 LABEL
820 PRECUT LABEL

82 BINDERS
830 GBC
831 3-RING NOTEBOOKS
832 ACCO

83 AUDIO/VISUAL MATERIAL
830 PROJECTION LAMPS
831 VIDEO TAPES
832 AUDIO TAPES
833 PHOTOGRAPHIC FILM
834 FILM CASES & REELS
835 SLIDE TRAYS
836 SLIDE MOUNTS
837 BATTERIES
838 GAFER TAPE
839 EASEL PADS/MARKERS
840 A/V PARTS & EQUIPMENT

84 EQUIPMENT RENTAL
890 A/V EQUIPMENT FEE
891 DuMONT VHS RENTAL
892 VIDEO/MONITOR RENTAL
893 LARGE SCREEN PROJECTION UNIT RENTAL
894 VIDEO ITEM (SINGLE) RENTAL

85 GRAPHIC ART MATERIAL
850 ART MATERIAL

86 FILM/VIDEO MATERIAL
860 PROJECTION LAMPS
861 VIDEO TAPES
862 AUDIO TAPES
863 PHOTOGRAPHIC FILM
864 FILM CASES AND REELS
865 SLIDE TRAYS
866 SLIDE MOUNTS
867 BATTERIES
868 GAFER TAPE
869 EASEL PADS/MARKERS
870 A/V PARTS & EQUIPMENT
871 PURCHASED PROGRAMS
872 STOCK PROGRAM MATERIAL
874 MAGNETIC FILM

87 MISCELLANEOUS MATERIALS
897 MISCELLANEOUS MATERIAL

98 IN-HOUSE DUPLICATION
981 BLANK VIDEO TAPE STOCK

OUTSIDE PURCHASED SERVICES (P)

55 PRESENTATION SLIDE CREATION
550 35mm
551 OVERHEAD

56 PRESENTATION MUSIC AND TALENT SERVICES
560 MUSIC & TALENT SERVICES

60 FILM/VIDEO RENTAL
600 EQUIPMENT RENTAL

61 FILM/VIDEO ANIMATION
610 ANIMATION

62 FILM/VIDEO EDITING
620 1" VIDEO EDIT FACILITY

63 FILM/VIDEO PHOTO PROCESSING
630 CORPORATE PHOTO LAB

64 FILM/VIDEO MAINTENANCE
640 EQUIPMENT REPAIR

65 FILM/VIDEO SLIDE CREATION
650 35mm

66 FILM/VIDEO SLIDE PROGRAMMING
660 MULTI-IMAGE

67 AUDIO/VISUAL MAINTENANCE
670 EQUIPMENT REPAIR

68 AUDIO/VISUAL VENDOR SUPPORT
680 EQUIPMENT RENTAL

69 FILM/VIDEO LIBRARY PROCESSING
690 PROGRAM DUPLICATION
691 VIDEO TAPE DUPLICATION
692 AUDIO TAPE DUPLICATION
693 35mm SLIDE DUPLICATION
694 16mm FILM DUPLICATION
695 VENDOR SUPPLIED MATERIAL

88 FILM/VIDEO FACILITIES/FEE
881 VIDEO EDITING ROOM FEE-PRIMARY
882 VIDEO EDITING ROOM FEE-SECONDARY
883 FILM EDITING ROOM FEE
884 SLIDE EDITING ROOM FEE
885 AUDIO EDITING ROOM FEE
886 MUSIC SELECTION ROOM FEE
887 DUBBING ROOM FEE
888 VIDEO STUDIO FEE
889 PRODUCTION EQUIPMENT FEE

89 FILM/VIDEO TRAVEL EXPENSES
880 TRAVEL EXPENSES

90 ART
901 VUGRPAHS
902 SIGNS, BANNERS, POSTERS
903 TECHNICAL ILLUSTRATION
904 KEYLINE
905 35mm SLIDES
906 OVERLOAD ARTISTS
907 ANIMATION ART
908 ARTWORK

91 COMPOSITION
911 TYPESETTING
912 KEYLINE (6912 ONLY)
913 VENDED FORMS

92 PRINTING
921 PRESS RUN
922 BINDERY
923 LAMINATING

93 LITHOGRAPHY
931 1/2 TONE NEGATIVES
932 SCREEN PRINTS
933 LINE NEGATIVES
934 4-C SEPARATIONS

94 PHOTOGRAPHY (6906/07)
941 PHOTOGRAPHY

95 FILM/VIDEO
953 TALENT SERVICES
954 SOUND SERVICES
955 LAB SERVICES
956 VENDOR SUPPLIED MATERIAL
957 SPECIAL EFFECTS
959 EDITING SERVICE
960 FILM RENTAL
961 CONSULTANT FEES
962 MESSENGER DELIVERY CHARGES
963 MUSIC RELEASE FEES
964 CAMERA OPERATOR
965 STILL PHOTOGRAPHER
966 VENDOR-RUSH CHARGES
967 PROGRAM DUPLICATION

96 DISPLAY BUILD
961 DISPLAY BUILD
962 SPACE
963 LOGISTICS
964 SUPPORT

97 PREMIUM/MODELS
971 PREMIUMS
972 MODELS
975 PREMIUMS (6994 ONLY)
976 MODELS (6904/05 ONLY)

99 FILM/VIDEO ADMINISTRATION FEE
990 COORDINATION

HS-596 (BACK) REV 7/83

Figure B.20: A Production Cost Record Including Outside Costs

Media Production Cost Record

AVC # _____

Program _____

Director _____   Client _____

Media _____   Company _____

Due Date _____   Department _____

Total Budget _____   Actual Internal Cost _____

Date Completed _____   Actual External Cost _____

Audiovisual Costs:

Production	Art:	In	Out	Photo:	In	Out	T.V.	Reproduction	Distribution

Production Outside Costs:

Vendor	Art	Purchase Requisition #	Date Rendered	Invoice Received/Signed	Type of Service

Figure B.21: A Time Sheet Showing Project and Non-Project Hours

	Project				Non-Project			
	Hours		Reg.	OT	Hours		Reg.	OT
	No:	Element			See Key			
MONDAY								
TUESDAY								
WEDNESDAY								
THURSDAY								
FRIDAY								
SAT/SUN								

Name:

Week of: No.

Overtime hours total:

Project key

Number	Title

Non-Project key
1. Auditions
2. Tours
3. Previews
4. Training others
5. Being Trained
6. Physical plant operations
7. Covering for other staff
8. Meeting setups
9. NEHO sound systems
10. Paper shuffling
11. Software loanouts
12. Equipment loanouts
13. Counseling
14. Budgets
15. Facilities cleanup
16. Planning
17. Reading
18. Working with vendors
19. Repairs
20. Clerical
21. Software research
22. Equipment research
23. Industry input
24. Field input
25. Administration

Figure B.22: A Freelancer's Time Sheet

FREELANCERS' TIME SHEET

Name: Billing Date: Rate:

Please itemize the following:

Day/days worked	Name of Projects	No. of hours

Supervisor:

Figure B.23: A Time Sheet for Charging Employee Time by Project

SPECIAL CODES

A	1	1ST SHIFT
	2	2ND SHIFT
B	0	REGULAR
	8	USER ALTERATION
	9	HOUSE ERROR
C	0	REGULAR TIME
	1	OVERTIME
	2	SUNDAY
	3	HOLIDAY

e.g. 181 = 1ST SHIFT, USER ALTERATION OVERTIME

COST CENTER

75 AUDIO/VISUAL CENTER

OPERATION

750 MEETING SUPPORT PLANNING
751 FACILITY COORDINATION
752 TRAVEL/DELIVERY
754 EQUIPMENT SET-UP/STRIKE
754 EQUIPMENT OPERATION
755 AUDIO/VISUAL CONSULTING
756 AUDIO/VISUAL FACILITY DESIGN
757 CUSTOMER EQUIPMENT REPAIR
758 EQUIPMENT LOAN-OUT
759 USER INCURRED WAIT TIME

NOTE: NON-CHARGEABLE TIME

1. LUNCHTIME: ENTER "L" ONLY IN JOB NUMBER

2. ENTER "N" IN JOB NUMBER
   ENTER SPECIAL CODE
   ENTER COST CENTER CODE
   ENTER CODE NUMBER FROM BELOW IN OPERATION NUMBER

900 VACATION
910 PERSONAL
920 ILLNESS
930 TRAINING & ADMINISTRATION
940 WAIT TIME
950 NON-CHARGEABLE CONSULTATION
960 MAINTENANCE
970 GENERAL & ADMINISTRATIVE
980 TEAM
990 SPECIAL PROJECTS
995 VENTURE PROJECTS

HT-820 REV 7/83

Figure B.24: An Individual Weekly Time Record Log

CORPORATE TELEVISION CENTER
TIME RECORD LOG: NAME: _____   , FOR THE PERIOD _____ THRU SUNDAY _____

	MONDAY	TUESDAY	WEDNESDAY	THURSDAY	FRIDAY	SATURDAY	SUNDAY	REMARKS
A. TV PRODUCTION #								
1. MEETINGS, GEN'L PREP.								
2. SCRIPTING								
3. VISUALS								
4. REMOTE SHOOTING								
5. STUDIO SET-UP/TEAR-DOWN								
6. STUDIO OPERATIONS								
7. CONTROL ROOM OP'NS								
8. DUBBING								
9. PREVIEW & EVALUATION								
10. OTHER								
B. GENERAL DISTRIBUTION								
C. LIBRARY DUPLICATION, ETC.								
D. MEETINGS								
E. REPORTS, OFFICE, ETC.								
F. EQUIPMENT, FACILITIES								
G. OTHER (TOURS, CLEANUP, TRAVEL)								
DAILY TOTALS								

Figure B.25: A Departmental Weekly Reporting Form

AUDIO VISUAL SERVICES DIVISION WEEKLY REPORT FORM

THIS REPORT IS DUE BY 10:00 AM ON MONDAY   .

NAME: _____   EMPLOYEE ID # _____

RECEIVED: _____

INPUT: _____

WEEK OF: _____

STAFF TIME

Project	Division	Date	Emp ID	Work Code	Type C/N/A	Hrs/ Days	Client Name	Comments

FACILITIES USAGE

Project	Division	Date	Emp ID	Work Code	Type C/N/A	Hrs/ Days	Client Name	Comments

MATERIALS/SUPPLIES USAGE

Project	Division	Date	Emp ID	Work Code	Type C/N/A	Units	Prem.	Client Name	Comments

DIRECT EXPENSES

Project	Division	Date	Emp ID	Work Code	Type C/N/A	Units	Cost/Unit	Client Name	Comments
83012	073			996	N			COMPANY TIME	

****** PLEASE NOTE THAT PROJECT #, DIVISION CODE AND EMPLOYEE ID # MUST BE FILLED IN FOR ALL ENTRIES ************************

Appendix C: A Directory of Expenses by Category

The following list does not include every possible item in every category, nor will the breakdown of the categories be common to all organizations. However, the list should give the reader a general overview of the many kinds of expenses and should help lead to a complete analysis of the specific expense items that apply to the reader's own situation and conditions.

SECTION A: INTERNAL, OR INDIRECT, EXPENSES

Internal, or indirect, expenses are those that, in most circumstances, are annual and ongoing. They are incurred regardless of the amount of work done by the organization and cannot be directly attributed to or allocated to specific projects.

Salaries and Benefits

- Regular staff salaries
- Regular staff overtime
- Bonuses and incentive payments
- Part-time/temporary help (e.g., temporary clerical help)
- Employee benefits (e.g., life, health and dental insurance, retirement, investment plans, meal subsidies, disability and sick pay, maternity leave)

Rent and Utilities

- Space rent *(Different spaces—such as storage space and office space—may have different rent rates.)*
- Heat and air conditioning, if not included in rent
- Building and maintenance, over and above what might be included in rent rate (e.g., special maintenance work on electrical and air conditioning systems)
- Office renovations and redecorating

Telephone

- Monthly instrument rental
- Local and long distance charges

- Special equipment or service charges (e.g., WATS lines, speakerphones, data sets, bridges, call forwarding)
- Installation and removal of telephone lines

Travel

- Air fares
- Hotel and motel
- Ground transportation, including mileage for use of personal vehicles
- Meal expenses
- Miscellaneous personal expenses (e.g., tips, telephone and valet services)
- Entertainment expenses

Office Equipment Purchases and Rentals

- Furniture, if not included in space rentals
- Typewriters, adding machines, copy machines and per-copy duplicating charges

Data Processing Equipment and Services

- Purchase or rental of computer terminals or personal computers
- Mainframe computer time and storage charges
- Line charges for specific computer networks, such as CompuServe
- Prepackaged computer software
- Specialized computer programming services

Office Supplies and Printing

- General office supplies (e.g., paper, pens, pencils, etc.)
- Special supplies (e.g., computer disks and paper, calendars, control boards, plus other items that might be general in nature but are specifically needed for AV functions, such as preprinted audio and videotape labels)
- Special printing (e.g., custom labels and forms, script paper, letterheads)

Books and Periodicals

- General publications (newspapers and business magazines)
- Trade publications *(Many of these are available without charge.)*

Special Employee Awards

- Tuition refund plans for additional education

- Cash awards for completing certain educational programs related to the business of your corporation, e.g., passing Securities Exchange Commission (SEC) licensing requirements
- Matching gifts to higher education programs

Education and Training

- Membership dues in professional organizations
- Conference, workshop and seminar registration fees *(Note that the cost of travel to these events may be charged to this account rather than to the regular travel budget.)*

Shipping and Mailing

- Regular and special postage *(Note that an internal mail department may levy additional charges for handling priority and registered mail.)*
- Express and shipping, including overnight courier services, shipping equipment to remote locations and airline overweight charges
- Local messenger and delivery services

Professional Fees and Services

- Special consultants (e.g., for the design of production facilities or for feasibility studies)
- Legal and accounting services
- Computer programming services, if not included under data processing charges

Company Overhead Allocations

- Indirect and lumped charges for all other internal service departments (e.g., law, purchasing, personnel, and perhaps even a share of the salaries of company executives and their support staffs)

Equipment Depreciation and Maintenance

- Annualized depreciation on audio visual equipment owned and operated by the company
- Repairs and maintenance on that equipment

Annual Equipment Lease Costs

Insurance

- Special coverage for equipment when traveling *(Usually the equipment will be covered by the company's blanket policy while it is in the home office.)*
- Liability coverage for crew, talent and visitors to the AV facility
- Travel and liability insurance for outside crew and talent while traveling
- Liability coverage on the premises and property of other individuals and companies, when working on location *(Generally other companies and government agencies will require proof of such insurance before issuing a permit to work on their premises.)*

SECTION B: INTERNAL DIRECT EXPENSES

Internal direct expenses are, in most circumstances, indirectly project related. While these are items usually purchased in bulk, they are used for specific projects and can be allocated to that project after the fact.

Audio and Videotape

Film

- Still, motion picture, Polaroid, and printing paper and chemicals

Film Processing

- Internal and external

Graphics Materials

- Storyboard pads, photostat supplies and other specialized materials

Printing

- Company standard labels and survey forms

Storage

- Master tape and film materials in off premises libraries

SECTION C: EXTERNAL DIRECT EXPENSES

External direct expenses are incurred directly as the result of work on a specific project and can be so identified and charged to that project.

Personnel Costs

- Professional talent, including payments to agents, talent union pension and welfare funds, and other required payments (e.g., meals, travel, costume costs)
- Casting expenses, including casting agents fees and audition studio rentals
- Professional services (e.g., lighting and set designers, photographers, artists, film and video editors, cinematographers and multi-image programmers, including their travel and expenses)
- Payments to location scouts

Travel Costs

- Specific air fare, hotel and per diem expenses for staff in connection with location scouting and location photography, production or staging
- Local travel for research or auditioning talent or to and from local outside production studios

Licenses

- Music library needle drop fees
- Commercial music licenses
- Stock photography licenses

Equipment and Facilities Rentals

- Rental of special equipment for specific projects (e.g., lighting equipment, cameras, recorders)
- Rental of outside production and post-production facilities
- Rental payments for locations (e.g., homes, stores, offices)

Props and Sets

- Building and painting sets
- Purchase or rental of furniture or special props
- Construction of special props or the printing of special forms

Special Staging Costs

- Stage crews and lighting personnel for conference staging
- Conference equipment rented through a staging house
- Construction of stage sets or stage draping

Security Guards

- Watching equipment on location

Duplication and Shipping

- Audio and videocassette duplication costs, including stock
- Printing of labels
- Slide duplication, mounting, collation and traying
- Bulk or individual shipping of duplicated materials, including shipping master materials to duplicator and the cost of shipping packaging

Appendix D:
The AVMA Survey

Preparing this book required gathering background information about the budgeting practices of some typical audio visual managers. For that purpose, a survey was prepared and mailed to the membership of the Audio Visual Management Association (AVMA).

The AVMA is an international organization for managers of AV functions in business and industry. The goal of the organization is to improve the overall effectiveness of its members as they deal with their company's employees, customers, shareholders and the general public through the production, distribution and use of all audio visual media.

Membership in the AVMA is limited to a small (135 maximum), select group of professional men and women actively engaged in the management of corporate AV communications. At the time of the survey, there were 120 active members of the organization. The members of the organization, while representing only a small sample, do cover a broad spectrum of company sizes, industries, geographical territories and organizational structures. To qualify for membership the individual must have a managerial responsibility. Furthermore, only one individual from any one corporation or subsidiary may be a member at any one time.

Members represent a variety of companies, including giants like AT&T, IBM and Exxon as well as a number of medium-sized firms such as Fisher Scientific, Garrett Airesearch, Foxboro and Eaton Corp. There are also small companies such as Ohio Medical Products, Rhode Island Hospital Trust National Bank and Alabama Cast Iron Pipe Company.

Thus, for the purposes of the survey of this group is ideal. Because of the membership limitations, duplicate responses were eliminated. We were also assured of a response that represented the entire audio visual service department rather than just one media or service. Finally, meaningful comparisons were possible because all members must work for business or industry; there are no manufacturers of audio visual equipment, independent producers or freelancers, or government agencies in the response sample.

Sixty-nine valid responses were received, representing a 57.5% response rate. A copy of the survey is shown in Figure D.1.

DEMOGRAPHICS

Part A of the survey was intended to establish the demographics of the survey group. Not all of the responses are important for the purposes of this book, but the following may be of interest.

A.1: Primary Business

Manufacturing	41%
Financial services	16%
Transportation	3%
Health care	7%
Communications	9%
Petroleum products/services	7%
Utilities	6%
Data processing equipment/services	4%
Sales/distribution/retail	6%
Other	1%

A.2: Number of Company Employees (Company Size)

Under 5,000	19%
5,000 to 20,000	35%
20,000 to 50,000	28%
Over 50,000	18%

A.3: Subsidiary Status

19% of the respondents work for subsidiary organizations.

ORGANIZATIONAL STRUCTURE AND 1983 BUDGETS

Part B deals with the organizational structure and budgets for the current year (1983).

Figure D.1: The AVMA Audio Visual Budgeting Questionnaire

AUDIO VISUAL BUDGETING QUESTIONNAIRE

A. Your Company
1. What is your company's primary business?   2. Number of Employees?
   a. Manufacturing ____
   b. Financial Services ____   a. Under 5,000 ____
   c. Transportation ____   b. 5,000 to 20,000 ____
   d. Health Care ____   c. 20,000 to 50,000 ____
   e. Communications ____   d. Over 50,000 ____
   f. Petroleum Products/Services ____
   g. Utilities ____
   h. Data Processing Equip/Svcs. ____
   i. Sales/Distribution/Retail ____
   j. Other _____ ____

3. Your company is ___; is not ___ a subsidiary of another corporation?
   (If your status will change on 1/1/84, indicate what it will be)

4. Do other departments at your company location do audio visual work? ___
   If so, how many?_____.

5. Throughout your corporation, including divisions and subsidaries, how
   many separate groups perform audio visual work? _____.

B. Your Department/Division
1. To what functional organization is your audio visual staff attached?
   a. Communications/Public Relations ____
   b. Personnel ____
   c. Corporate Training ____
   d. Corporate Administration ____
   e. Marketing/Sales ____
   f. Other _____ ____

2. Number of full time people on your AV staff? _____; Number of
   permanent part time?_____.

3. What is your total 1983 Annual Budget?
   a. Internal expenses, salaries and overhead $_____
   b. External expenses, materials and services $_____
   c. Other _____ $_____
                                        TOTAL $_____
   d. Capital Expenses, Equipment $_____
   e. Capital Expenses, Construction $_____

4. Is your 1983 operating budget greater than ___; the same as ___; or less
   than ___; your 1982 budget? By what percent different? _____%

5. Do you expect your 1984 budget to be greater than ___; the same as ___;
   or less than ____; your 1983 budget? By what percent different?_____%

NOTE: If any of the information requested is confidential or you would prefer not
   to put it in writing, but would be willing to discuss it with me by phone,
   please check here _____, and put your phone number on the next page.

(continued on next page)

Figure D.1: The AVMA Audio Visual Budgeting Questionnaire (cont.)

BUDGETING QUESTIONNAIRE   REVD AD-11

C. Your Budgeting System.   AVMA - 2

   1. What type of budgeting system do you now have?
     a. Annual budget, including overhead and expenses established by the
        company and not allocated to users ____.
     b. Operating budget set by company as in "a", with external costs charged
        directly to or paid directly by users ____.
     c. All operating and external expenses allocated to users on basis of per-
        centage of work performed?____.
     d. Operating expenses allocated as in "c", with external costs charged to
        or paid directly by users ___.
     e. Charge back system with all expenses charged to clients based upon
        established rates?____.
     f. Profit Center with rate card charges to internal and/or external users
        calculated to show a profit?_____.
     g. Other _____.

   2. If you allocate or charge operating expenses to users in any way, do those
      charges include: Equipment Depreciation ___; Floor space rental _____;
      Equipment maintenance _____; Your share of general company overhead _____.

   3. Do user departments have the option of buying audio visual services on the
      outside, without going through your department?_____.

   4. Has your budgeting system changed? If so, in what year ?_____. Please in-
      dicate by letter from "1, a-f", what it was before._____

   5. Is your present budgeting system: Unique to your organization ?_____;
      Similar to some other Departments? _____; Same as rest of company _____.

   6. How was your present budgeting system established? By executive order?___;
      By executive order at your suggestion? ____; By a change of company account-
      ing procedures? _____ Other?_____.

   7. Under which of the budgeting systems listed under "1" would you prefer
      to operate? _____.

   8. Do you presently sell facilities time ___ and/or services ___ outside your
      company formally ____; informally _____?

D. Please add any comments you would care to share_____

  _____

  _____

  _____

E. Please include in the return mailing, any budgeting forms you would care to share.

Name (Optional)_____ Company _____

Phone # _____ _____

B.1: Functional Organization to Which AV Staff is Attached

Communications/public relations (includes advertising, creative services, etc.)	59%
Personnel	3%
Corporate training	6%
Corporate administration	13%
Marketing/sales	13%
Other	6%

B.2: Full-Time Staff

55 or more	2%
20 to 54	14%
10 to 19	23%
6 to 9	33%
2 to 5	23%
1	5%

Average staff size = 12

B.2: Part-Time Staff

Forty-three percent employ part-time audio visual personnel. The largest employs 10 (this same respondent also employs 10 full-time personnel). The average respondent who employs part-time personnel has 2.7 positions.

B.3: Total 1983 Budget

Not all of the respondents were willing (or perhaps they were not able) to provide the requested breakdown of budgeting information. However, for those who did so, the figures were as follows:

Average internal budget	$515,000
Average external budget	320,000
Average reported capital budget	144,000

Only 14% had construction budgets in 1983, averaging $344,000.

For all respondents as a group:

	1983 Operating Budget	1983 Capital Budget
High	$3,000,000	$3,677,605
Low	90,000	5,000
Average	821,970	205,500
Median	540,581	75,000

Fully 34% of the respondents had operating budgets of $1 million or more.

An attempt was made to correlate operating budgets by size of company, and the results are shown in Table D.1. While it can be seen that the size of the audio visual budget generally reflects the size of the company, this is not always the case. No companies with fewer than 5000 employees had budgets over $2 million, and no companies with more than 50,000 employees had budgets under $250,000. However, it is also worth noting that fully 10% of the companies with fewer than 5000 employees did have budgets in excess of $1 million, and 22% of the companies with more than 50,000 employees had budgets of less than $500,000.

Table D.1: Correlation of Size of Annual Operating Budget to Size of Company

Company Size Number of Employees	Annual Operating Budget (in thousands) Under $250	$250-$500	$500-$1,000	$1,000-$2,000	Over $2,000
Under 5,000	36%	36%	18%	10%	-0-
5,000-20,000	33%	24%	19%	5%	19%
20,000-50,000	16%	21%	21%	37%	5%
Over 50,000	-0-	22%	12%	33%	33%

Because of the relatively small size of the sample, it is not valid to attempt to correlate size of budget either to type of industry or to the organization to which the audio visual department is attached.

B.4: Increase or Decrease in 1983 Budget as Compared to 1982

Sixty-seven percent indicated an increase in the 1983 budgets as compared to 1982, while 21% indicated a decrease and 12% indicated no change. As was pointed out in Chapter 1, "no change" is tantamount to a decrease, due to inflationary pressures. Therefore, a total of 33% can be assumed to have tightened their belts in 1983. The average increase was 13%, while the average decrease was 11%. However, of the total indicating an increase in 1983 budgets, 11% of those were held to increases of 5% or less.

B.5: Increase or Decrease in 1984 Budget as Compared to 1983

Sixty percent indicated that 1984 budgets would be greater than those granted in 1983; however, 13% of that number were held to increases of 5% or less. The average 1984 increase was 12%, slightly less than the 1983 increase. Another 16% anticipated decreases in their 1984 budgets averaging 15% (somewhat greater than the average 1983 decrease), while the remaining 24% projected no change in 1984 budgets as compared to 1983.

Summary

If we assume that a 5% increase is equivalent to no change due to inflation in both years, we can then plot the trend in audio visual budget increases as follows:

	Substantive Increase	Decrease or No Effective Increase
1982/1983	61%	39%
1983/1984	53%	47%

Thus, while a slim majority of audio visual organizations continue to receive annual budget increases, the gap is rapidly narrowing.

BUDGETING AND COST ALLOCATION SYSTEMS

In Part C we asked the respondents to attempt to identify the types of budgeting and cost allocation systems in place. The results are summarized in detail in Chapter 2, but it is worth repeating that any attempt to formally classify budgeting systems is bound to uncover a large number of systems that defy classification.

In Chapter 2 we listed five types of budgeting systems. For the purposes of the survey we identified six types, making a distinction between two different types of full allocation systems (categories C and D in question C.1). In both types, all internal and external expenses are allocated back to users or other areas of the company, but by two different methods. The responses by specific category and the varying combination systems are outlined below.

C.1: Budgeting System Currently in Place

A. Annual budget, no allocation (full overhead)	19%
B. Annual budget/external charged direct (direct cost)	30%
C. Operating and external expenses allocated	6%
D. Operating expenses allocated, external expenses charged direct	10%
E. Chargeback	23%
F. Profit center	3%
Combined systems (different clients treated differently)	
A and B combined	3%
A and E combined	1.5%
B and E combined	1.5%
C and D combined (included in full cost allocation for overall tabulation)	3%

Note: Categories C and D are combined in the full cost allocation group in the summary tabulations.

C.2: Overhead Expense Line Items Included in Calculations

Respondents were asked to identify which of the listed expense items were included in their calculations for full allocation and chargeback/profit center systems. The responses represent 46% of the total survey group.

Equipment depreciation	43%
Floor space rental	43%
Equipment maintenance	49%
Share of company overhead	39%

In addition, 4% of the respondents include the costs of equipment leases in determining their allocations or charge rates. Only 5% of the total group include all four categories in establishing their charges, including only one respondent in the chargeback and profit center category.

Summary

It is obvious that none of the listed categories, all of which are essential to the operation of an audio visual department, is included in the calculations for even 50% of the companies. Our question is: How accurate can any allocation or chargeback system be if these factors are not included? This would certainly apply for the first three items. In some cases, companies may not have a system which would allow an allocation of general company overhead back to the operating division.

C.3: Outside Competition

This question was designed to find out how much competition audio visual departments have to face from outside vendors. Could user departments go directly to outside audio visual sources, or must they go to the in-house AV department first? We broke the responses down by general type of allocation/budgeting systems to see what difference the kind of system made to these responses. A "no" response indicates that user departments are directed to go to the in-house AV department first. "Yes" indicates that users may go directly to outside sources, bypassing the in-house AV department.

Type of System	Yes	No
Full overhead	63%	37%
All allocation systems	77%	33%
Chargeback and profit center	82%	18%
Total, all responses	75%	25%

Summary

I was quite surprised by the results. It was expected that most companies would require that in-house services be the first choice. Quite the opposite is true. I also expected that company policy would offer the most protection for the chargeback and profit center operations; again, exactly the opposite is true. To make the situation even worse, many of those who responded "no" indicated that while that may be company policy, many users went directly outside anyway. The large percentage of those in full overhead systems where users can go directly outside doesn't even make sense, since under those systems the services are "free." I wonder how long company centralized computer functions could continue to function if such policies existed in that area.

C.4: Changes in Budgeting Systems

In this question, respondents were asked to indicate changes in budgeting systems. No time limit was specified, so some systems may have changed as late as 1982 (the year before the survey). As to projections for future changes, most indicated that these changes would take place in 1984.

Type of Budgeting System	% Previous	as of 12/31/83	% Future
Full overhead	34%	19%	17%
Direct cost allocation	28%	30%	28%
Full cost allocation	18%	19%	19%
Chargeback system	12%	23%	25%
Profit center	1.5%	3%	5%
Combination systems			
Allocation and overhead	4%	3%	3%
Allocation and chargeback	1.5%	3%	3%

Summary

While only 66% of the users of audio visual services were being charged for some or all of those services only a few years ago, by as early as 1984 some 83% will find themselves paying directly or indirectly for audio visual services and materials. I do believe, however, that this is only a reflection of a general trend toward cost-accountability in business and industry and is not specifically directed at the audio visual function per se. This is evident from the next question.

C.5: Commonality of Budgeting/Allocation Systems Within the Company

Respondents were asked to indicate whether the budgeting system currently in use was unique to the audio visual organization, similar to systems in use by some other organizations or common throughout the company. Again, in correlating the results we used the three primary types of systems.

Type of System	Unique	Similar	Common
Full overhead	-0-	13%	87%
All allocation systems	21%	46%	33%
Chargeback and profit center	12%	76%	12%
Total survey	14%	46%	40%

Note: Combination systems were included in their primary allocation category, either full overhead or one or another allocation category.

Summary

I expected the responses for full overhead and allocation systems. What is surprising is the response for chargeback and profit center systems. I anticipated that the vast majority would be unique, but it turns out that only a small number are unique, and an equal number are common throughout the company. In the largest number of cases, there is a similar system in place in at least one other department within the company. Again, this may be a general reflection of the increasing awareness of the need to identify and control expenses.

C.6: How Was the Budgeting System Established?

The primary purpose of this question was to find out just how much input the AV department itself may have had in deciding what kind of budgeting/allocation system was going to be established.

Type of System	Executive Order	AV Department Input	Company Policy or Procedure
Full overhead	72%	7%	21%
All allocation systems	31%	25%	44%
Chargeback and profit center	31%	31%	38%
Total survey	40%	23%	37%

Summary

It is certainly encouraging to see how much influence the AV departments have had in establishing their own financial operating systems—even for those few who had input to full overhead systems, which are generally decided by executive direction or company practice. Actually, I had expected that almost all of these would have been based on policy and procedure and am surprised at the incidence of responses indicating executive input to the system. The response with regard to chargeback and profit centers appears to be just what I would have expected. Many of the respondents noted that their systems were established some time back in "ancient history." One indicated that the chargeback system had been established as the result of a company audit.

C.7: Preferred Budgeting/Allocation System

Sixty-seven percent of the respondents indicated that they were happy with the budgeting system they now had. The other 23% said they would like to have a different system. The preferred systems are shown, with the percentages of managers preferring that system.

Type of System	Selected by Those Preferring a Different System
Full overhead	22%
Direct cost allocation	26%
Full cost allocation	9%
Chargeback system	31%
Profit center	4%
Combination systems	
Allocation and overhead	4%
Allocation and chargeback	4%

Summary

Surprisingly, 31% indicated a preference for the chargeback system, and 4% indicated a preference for profit center. The small number (9%) who selected full cost allocation suggests that those who want to make a change want to make it all the way.

C.8: Sales of Facilities or Services Outside

Because many companies are now renting their production facilities and producing programs for outside clients, this question was designed to find out to what extent this practice had spread, and whether or not the type of budgeting/allocation system had an impact on this. It also asked whether this was done on a formal or informal basis. The results are shown in Table D.2.

Table D.2: Audio Visual Departments Doing Work for Clients Outside Their Companies

Type of System	Basis	Rent Facilities Only	Sell Services Only	Both Facilities and Services	Total
Full overhead	Informal		7%	26%	33%
	Formal				-0-
All allocation	Informal		9%	15%	24%
	Formal			3%	3%
Subtotal for group		9%	18%	27%	
Chargeback and	Informal	12%	6%	29%	47%
profit center	Formal	6%		12%	18%
Subtotal for group		18%	6%	41%	65%
All systems	Informal	3%	8%	21%	32%
combined	Formal	1.5%		4.5%	6%
Total for all		4.5%	8%	25.5%	38%

Summary

Obviously, the practice is widespread, even among those departments operating on a full overhead basis. It is hardly surprising that fully 65% of the departments on a chargeback or profit center basis do business on the outside, though only 18% do so formally. I have to wonder what the reaction of commercial producers will be to this.

IN CONCLUSION

This survey has proven invaluable to the basic work involved in preparing this book, and I am indebted to the Audio Visual Management Association and its membership for assistance in gathering the data. However, this database constitutes only an initial step in what I hope will be an ongoing research process.

Index

About the Author

Richard E. Van Deusen is manager, audio visual communications for the Prudential Insurance Co. He was responsible for installing Prudential's first television production facility and for initiating a pilot program using video for sales skills training. He also supervised the installation of Prudential's 700-unit video network.

Mr. Van Deusen is widely known as an authority on audio visual cost allocation systems, particularly the chargeback system, and is a frequent lecturer and panelist on the subject. He is also the author of several articles in professional magazines. He has served as a board member and officer of the Audio Visual Management Association and the International Television Association, of which he is a founder.

Mr. Van Deusen holds a B.A. from Dickinson College and a Master of Fine Arts from Boston University.

**Other Titles Available from
Knowledge Industry Publications, Inc.**

Managing the Corporate Media Center
by Eugene Marlow
ISBN 0-914236-68-7   hardcover $29.95

Video User's Handbook
by Peter Utz
ISBN 0-86729-036-6   hardcover $24.95

The Nonbroadcast Television Writer's Handbook
by William Van Nostran
ISBN 0-914236-82-2   hardcover $29.95

Professional Guide to Video Production
by Ingrid Wiegand   (forthcomimg)
ISBN 0-86729-067-6   hardcover $39.95

Video Editing and Post-Production: A Professional Guide
by Gary H. Anderson
ISBN 0-86729-070-6   hardcover $34.95

The Handbook of Interactive Video
edited by Steve Floyd and Beth Floyd
ISBN 0-86729-019-6   hardcover $34.95

Video in Health
edited by L. George Van Son
ISBN 0-914236-69-5   hardcover $29.95

A Practical Guide to Interactive Video Design
by Nicholas V. Iuppa
ISBN 0-86729-041-2   hardcover $34.95

**The Video Age: Television Technology and Applications
in the 1980s**
ISBN 0-86729-033-1   hardcover $29.95

**Knowledge Industry Publications, Inc.
701 Westchester Ave.
White Plains, NY 10604**

We intend to continue to monitor audio visual budgeting trends and systems. If you would be interested in participating in surveys similar to the AVMA survey, please send us your name and address.

Name _____ Title_____

Company_____

Address _____

City _____ State_____Zip_____

Return to: Knowledge Industry Publications, Inc.
            701 Westchester Avenue, White Plains, NY 10604
            Attn: Ellen Lazer, Senior Editor

Your reactions to the level and content of this book will be of value. Please feel free to comment below:

• Is there any aspect of budgeting that you would like to see covered in more detail?

_____

_____

_____

• Do you have a particular budgeting technique that was not covered but which you feel would be of interest to others?

_____

_____

(If so, please complete the top part of this questionnaire and return it with this portion.)

Please return to: Knowledge Industry Publications, Inc.
                  701 Westchester Avenue, White Plains, NY 10604
                  Attn: Ellen Lazer, Senior Editor